Patrick Moore's Practical Astronomy Series

D1160979

Other Titles in this Series

Lunar and Planetary Webcam User's Guide

Martin Mobberley

With 153 Figures

Springer

Martin Mobberley
Denmara
Old Hall Lane
Cockfield
Bury St Edmunds
Suffolk UK
IP30 0LQ
martin.mobberley@btinternet.com

British Library Cataloguing in Publication Data
A catalogue record for this book is available from the British Library

Library of Congress Control Number: 2006921111

Patrick Moore's Practical Astronomy Series ISSN 1617-7185
ISBN-10:1-84628-197-0
ISBN-13:978-1-84628-197-6
eISBN:1-84628-199-7

Printed on acid-free paper.

Printed in Singapore.   SPI/KYO

9 8 7 6 5 4 3 2 1

Springer Science+Business Media
springer.com

*To Damian Peach, Dave Tyler, Jamie Cooper, and Mike Brown, whose lunar
and planetary imaging enthusiasm and daily e-mails spurred
me on to write this book.*

Preface

We Live in Exciting Times

I have been looking at the moon since 1968 when I acquired a humble childhood refractor made by Prinz. It was a 30-mm aperture instrument with a 10–30× zoom feature. I still have that telescope today and it still works well. Over the years I acquired bigger and bigger instruments and tried some lunar and planetary photography. But I could always see far more visually than I could photograph because the eye and brain are a remarkable combination and far superior to photographic film, at least for studying the moon and planets. In 1985 I caused much excitement at BAA (British Astronomical Association) Lunar Section meetings when I showed some videotapes I had taken with an experimental English Electric Valve CCD camera I had used to videotape the moon. Here was a quantum leap in imaging: something that could rapidly study the lunar surface and record details as well as the eye, at least on the bright, high-contrast moon. However, as the 21st century dawned, even newer technology came to the fore. It was awesome in its power, but affordable too: surely, a unique combination for amateur astronomy. The technology was the humble USB webcam, combined with an incredible software package called Registax, developed by Cor Berrevoets. The technical combination of a webcam with image stacking and processing tools, controlled by a modern PC is truly staggering. Any amateur can now take back-garden planetary images that reveal more than the human eye and would have been the envy of professional observatories only 10 years ago. I cannot remember a more exciting time to be in amateur astronomy. I hope this book may inspire a few more imagers to emerge and help breed the next generation of lunar and planetary imagers. It is now possible for anyone with a modest backyard telescope to capture stunning images and contribute real scientific observations. If this book helps produce just one keen planetary imager, I will feel it has been worth it.

Martin Mobberley
Suffolk, United Kingdom
December 2005

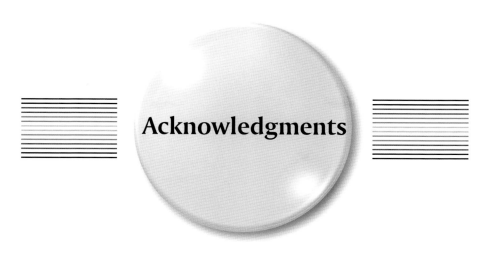

Acknowledgments

As was the case with my previous two Springer books, I am indebted to the outstanding amateur astronomers who have donated images to this new work. I feel humbled that such great names as Isao Miyazaki, Damian Peach, and Eric Ng (to name just three) have been so willing to share their images with others via my Webcam User's Guide. I am especially grateful to Damian as he has donated more images than anyone else to this work and it is only by his example, from the cloudy UK, that I was coaxed back to lunar and planetary imaging in the first place. While writing this work I was encouraged, on a daily basis, by e-mails from Damian, Dave Tyler, Mike Brown and Jamie Cooper as we all shared our nightly experiences regarding our webcams, telescopes, and the vagaries of the atmosphere and "seeing." So my special thanks go to those four as well as to David Graham, Mario Frassati, and Paolo Tanga, whose excellent Mercury and Mars maps are reproduced in these pages. The Frassati/Tanga Mars map is the most useful map for the webcam/visual user that I know of; it is superb, with every feature named but without being cluttered. But perhaps all our thanks should go to software genius Cor Berrevoets. Without Cor's Registax freeware, using webcams for planetary imaging would probably never have caught on at all.

In alphabetical order, the help of the following lunar, planetary, and image processing experts is gratefully acknowledged: Cor Berrevoets, Mike Brown, Celestron International, Antonio Cidadao, Jamie Cooper, Mario Frassati, Maurice Gavin, Ed Grafton, David Graham, Paolo Lazzarotti, Isao Miyazaki, NASA/JPL, Eric Ng, Gerald North, Donald Parker, Damian Peach, Christophe Pellier, Barry Pemberton (Orion Optics), Maurizio Di Sciullo, Paolo Tanga, Dave Tyler, Unisys Corp., and Jody Wilson (Boston University).

I am also indebted to my father, Denys Mobberley, for his tireless help in all my observing projects; to John Watson, for carefully reading through the original manuscript; and to the production staff at Springer for all their help, especially Jenny Wolkowicki and Chris Coughlin.

Martin Mobberley
Suffolk, United Kingdom
December 2005

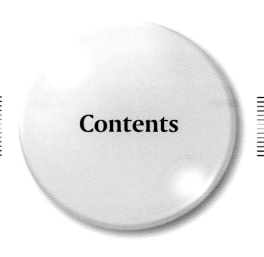

Contents

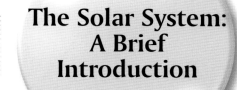

CHAPTER ONE

The Solar System: A Brief Introduction

This is primarily a book about imaging the planets with webcams and is aimed at amateur astronomers who already have some basic knowledge of the solar system. However, we all have to start somewhere and it is quite possible that a few complete beginners will be attracted to this book, with virtually no prior knowledge of astronomy at all. This chapter is for the total novice. If you are familiar with the structure of our solar system, feel free to skip this introduction. If not, then start your adventure here. I have deliberately written this introduction to be as simple as possible, so that almost anyone will (hopefully) understand it.

We live in a solar system (Figure 1.1) comprising one Sun, nine planets, hundreds of thousands of asteroids, and well over one thousand comets. If I sound a bit vague about exactly how many asteroids and comets there are, well, that is deliberate. Firstly, no one knows. Secondly, if you count objects no bigger than large boulders or snowballs as asteroids and comets, there must be millions. Figures 1.2 through 1.7 show some of the solar system's major bodies.

Billions of years ago, material in our solar system started to drift toward a common center of gravity. The vast bulk of the material, comprising mainly hydrogen, formed the Sun. As soon as enough mass was present, nuclear fusion started and the Sun began to shine, illuminating the early solar system and announcing its presence as a star in this part of our galaxy. All of the stars in the night sky are Suns; some are bigger, some are smaller, but all of them are a huge distance away.

In passing, I would like to stress early on that you must *never, ever* stare at the Sun, even with the naked eye, and certainly never with a telescope (unless the telescope has special full-aperture solar filters and you are an expert; see Chapter 16 for more details). The Sun is dangerously bright and permanent eye damage can easily occur. You have been warned!

The distances to other stars (and, therefore, other solar systems) are measured in light years. The closest star to us, Proxima Centauri, is 4.2 light years away.

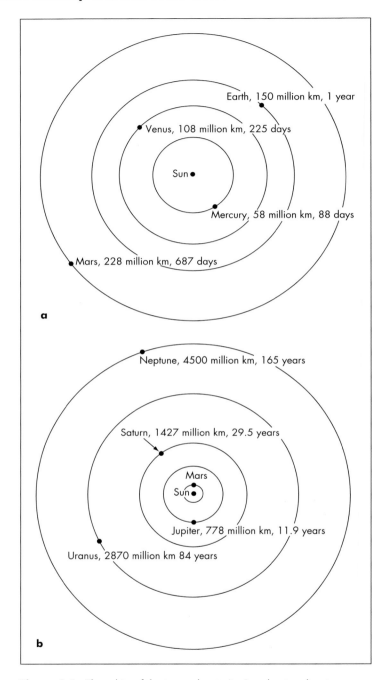

Figure 1.1. The orbits of the inner planets (top) and outer planets (bottom) are shown. Diagram: courtesy of Gerald North. Average solar distances are represented and Pluto is excluded due to space limitations.

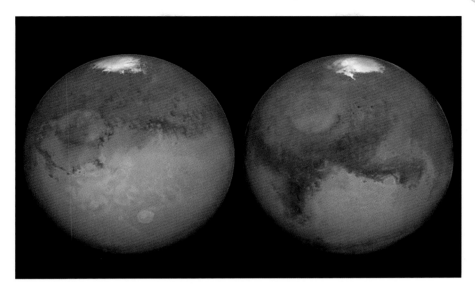

Figure 1.2. Mars, imaged at its closest to the Earth. Image taken by the Hubble space telescope on August 26, and 27, 2003. South is up. The Solis Lacus is shown on the left of the left-hand image; the Syrtis Major is shown on the left of the right-hand image.
Image: NASA/STScI.

Figure 1.3. The 33 × 13 km asteroid Eros, imaged by the NEAR/Shoemaker spaceprobe.
Image: NASA.

Figure 1.4. Comet Hyakutake from Tenerife on the night of March 24/25, 1996. Image: M. Mobberley.

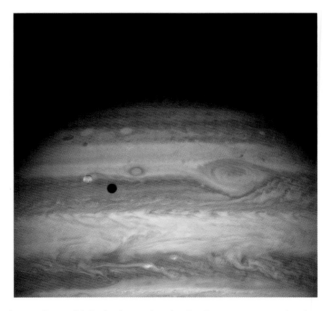

Figure 1.5. Jupiter, Io, and Io's shadow, taken by the Cassini space probe. Image: NASA.

Figure 1.6. Saturn imaged by Cassini. Image: NASA.

In other words, it would take a beam of light, traveling at 300,000 kilometers per second, 4.2 years to get from here to that star. By comparison, light takes just over a second to get from the Moon to the Earth and just over eight minutes to get from the Sun to the Earth. As nobody knows (yet) how to travel faster than light, it can be seen that traveling to other stars in a reasonable timescale is pretty well impossible with current technology. One only has to look at the stars in the sky and see

Figure 1.7. Uranus imaged by Voyager 2: Image: NASA.

how faint they are to appreciate how far away they are. Each one would be a blindingly brilliant Sun when seen close up.

In addition to the material that formed our Sun almost five billion years ago, the leftover chunks of heavy elements clustered together and became the planets and asteroids that we see today.

All of the planets in the solar system orbit the Sun in the same direction as the Sun rotates. But, whereas the Sun rotates in 25 days, the planets orbit the Sun in much longer periods. Looking from above the solar system almost everything orbits the Sun counterclockwise. The only real exceptions to this are the long-period comets, which can orbit the Sun in any direction and at any angle.

In order outward from the Sun, the major planets are Mercury, Venus, Earth, Mars, Jupiter, Saturn, Uranus, Neptune, and Pluto. Most (but not all) of the asteroids or minor planets live in the asteroid belt between Mars and Jupiter. These are worlds smaller than our Moon, but there are hundreds of thousands of the smaller (kilometer-sized) ones. As one moves inward from Mars to the Earth and closer to the Sun, we encounter asteroids closer to home, too. These are classified as Amor, Apollo, and Aten asteroids, and some NEOs (Near Earth Objects) are considered a threat to the Earth. The most dangerous ones are called PHAs, or Potentially Hazardous Asteroids.

Locating the Planets

Finding the brightest planets in the night sky, when they are at their best, is not difficult. But it does require a bit of experience in knowing where north, south, east, and west are, and in recognizing the brightest constellations.

The inner planets, Mercury and Venus, are only ever visible in twilight just after sunset (in the west) or just before sunrise (in the east). Actually, Venus can be visible in a dark sky, but I am trying to keep things simple. Mercury is a bright object, but so low down in twilight that it is often very tricky to locate. However, Venus is unmistakeable. If you see a brilliant object that looks like a plane, in a dawn or dusk sky, but it is stationary, that is almost certainly Venus. Apart from the Moon, Venus is the brightest object you will ever see in the night (or twilight) sky.

Moving out beyond the Earth, the planet Mars, when at its best, can be a brilliant object, too. It can never be as bright as Venus, but is always one of the brightest star-like objects in the sky and is distinctly red in color, a dead giveaway to its identity. There is a trick here to finding planets if you are a complete beginner. Bright stars tend to twinkle rapidly, but bright planets do not because they are extended objects, not just tiny pinpoints of light. If you are looking for a planet but are confused by the star patterns, but also know that one of those bright things in "that direction" is a planet, just check if it is twinkling. When Mars comes close to the Earth it is unmistakeably Mars: it is red, not twinkling, but dazzling. But Mars has bright years and dim years (depending on exactly how far away it is) and in the dim years it is just another bright star.

Jupiter, the giant planet of the solar system, is always bright and never twinkles. It is never as bright as Venus but it is usually brighter than Mars. Through binoculars or a small telescope, its largest "Galilean" Moons are obvious: a line of four dots leading out from the planet, some to the right, some to the left.

Next we come to the most spectacular planet in the night sky. Saturn, when its rings are fully open, is an awesome sight. However, despite its appearance through a telescope, it is only as bright as the brightest stars and locating it can be a bit confusing for the beginner. Again, I would recommend the twinkle test if you are confused. As soon as you point a telescope at Saturn you know you have it. Nothing else looks like it! The rings are detectable even at 20× magnification.

The remaining planets in the solar system are not visible with the naked eye. Uranus and Neptune can be seen with binoculars, but finding Pluto requires a decent telescope. Even in the modern era, detecting any features on Uranus and Neptune is a major challenge. In recent years, some astronomers have queried whether tiny Pluto should really be classed as a planet or a big asteroid, but most astronomers prefer to leave it classified as a planet.

Table 1.1 lists the most important facts for the Sun and the major planets in our solar system. I have included our own Moon in the table, as it is a major observing target but, of course, it orbits the Earth, whereas the nine planets orbit the Sun. I have also included the largest minor planet, Ceres, in the asteroid belt, as the asteroid "representative."

So how do you find out where those planets are in the night sky tonight? The best guides can be found in monthly magazines like *Sky & Telescope* or *Astronomy*, or, in the **U.K.**, *Astronomy Now* or the BBC *Sky at Night* magazine. These magazines all feature a circular centerfold pull-out that you hold above your head to simulate the night sky. For those who prefer software, there is a

Table 1.1. Facts about our solar system

Object	Avg. Distance from Sun (millions of km)	Closest to Earth (millions of km)	Orbital Period (Year)	Diameter (km)	Rotation Period
Sun	N/A	149.6 (av)	N/A	1,391,980	25.4 days
Mercury	58	80.0	88 days	4,878	58.6 days
Venus	108	38.3	225 days	12,104	243.0 days
Earth	149.6	N/A	365 days	12,756	24.0 hours
Moon	149.6	0.238	Orbits Earth	3,476	27.3 days
Mars	227.9	56.0	687 days	6,794	24.6 hours
Ceres	400	250	4.60 years	950	9.0 hours
Jupiter	778	590	11.86 years	142,884	9.9 hours
Saturn	1427	1200	29.42 years	120,536	10.2 hours
Uranus	2870	2584	84.01 years	51,118	17.2 hours
Neptune	4497	4306	164.80 years	50,538	16.1 hours
Pluto	5906	4296	247.70 years	2,324	6.4 hours

The average distance from the Sun is given, but the actual distance extremes can be quite significant. For example, Pluto can come closer to the Sun than Neptune, but, on average, it is much further away. Pluto is probably one of the largest of the Trans-Neptunian asteroid-like worlds (sometimes called Kuiper Belt objects). The table above does not list any comets, although many short-period ones (orbital periods under 200 years) are present in the inner solar system. Long-period comets can have orbital periods of thousands of years and so only appear once, before returning to deep space, well beyond Pluto.

huge choice of so-called planetarium packages that will bring the night sky to life on your computer screen. These have the added advantage that you can zoom in on planets and reveal objects like Jupiter's Moons. You can also access masses of extra information on planets, stars, and galaxies by clicking your mouse on the object. The most popular planetarium software packages are those made by Starry Night (www.starrynight.com) and there are various levels of sophistication up to Starry Night Pro 5.0. Other packages are worth looking into as well. My favorite is Guide 8.0, produced by Project Pluto (www.projectpluto.com), which is both powerful and easy to use. For real experts, The Sky, from Software Bisque (www.bisque.com) is the ultimate package and it can integrate with CCD images and control telescopes. Redshift 5 (www.focusmm.co.uk) and Skymap Pro (www.skymap.com) also have loyal followings. However, for absolute beginners, a copy of one of the monthly astronomy magazines is probably your best bet.

Space Probes And Hubble

As well as observing the planets out of sheer fascination and actually seeing them with your own eyes, the images returned by planetary space probes throughout the last few decades have been truly staggering and are well worth tracking down on the web. While manned spacecraft have only traveled as far as the Moon, unmanned robotic spacecraft have traveled to every planet except Pluto. Traveling at speeds as fast as 60,000 kilometers per hour (17 kilometers per second), these probes still take years to arrive at the most distant planets, Saturn, Uranus, and Neptune, indicating just how large our solar system is and how slow our rockets are for interplanetary travel! We also have the Hubble Space Telescope images of the planets and ground-based images from professional observatories. This might all lead one to think that backyard images of the planets are of little scientific use, but nothing could be further from the truth. Space probes have a limited lifetime and a specific set of tasks to perform. It is very rare for any space probe to constantly monitor the whole surface of a planet in the way that an amateur astronomer (or network of amateur astronomers) can. Likewise, the Hubble space telescope has only maintained an occasional coverage of the planets and mainly when they have been at their best. Strange though it may seem, dedicated networks of amateur astronomers are best placed to study the planets, as even ground-based professional observatories tend to concentrate on imaging the most distant objects. Even today, if a dust storm erupts on Mars or two spots start to merge on Jupiter, it is the amateur astronomers who will be collecting the most images. So from your own backyard you can do real science, if you want to.

Of course, to make any serious observations you really need a decent telescope and, ideally, one of about 200 mm aperture or larger. For webcam work you need a PC and a webcam. The only other thing you need is enthusiasm and this book. I would like to think that there is enough information contained in these pages to really inspire a few people to become serious planetary imagers.

Good luck on your journey, but a word of warning. This hobby can become very addictive: you have been warned!

The opposition dates and sizes of the planets Mars, Jupiter, and Saturn are listed below. These are the best times to view these planets. An outer planet is at opposition when it is directly opposite the Sun in the sky. This occurs when it is closest to the Earth and it is crossing the meridian (i.e., is highest in the sky) at midnight.

MARS: 2007 Dec 24; 2010 Jan 29; 2012 Mar 3.

JUPITER: 2006 May 4; 2007 Jun 6; 2008 Jul 9; 2009 Aug 14; 2010 Sep 21; 2011 Oct 29; 2012 Dec 3.

SATURN: 2006 Jan 27; 2007 Feb 10; 2008 Feb 24; 2009 Mar 8; 2010 Mar 22; 2011 Apr 4; 2012 Apr 15.

Webcams, Plus a "Quick Start" Guide

This book is about observing the Moon and planets using webcams, those small, cheap little cameras that plug into your PC for live video messaging. If you are comfortable with PCs, software, and telescopes and want to get started imaging *now*, read the Quick Start Guide at the end of this section first. However, if you are less confident with imaging technology, but maybe fairly knowledgeable about telescopes, you will need to read on a bit further and, at some point, jump to the full webcam beginner's guide in Chapter 7. If you are new to astronomy *and* imaging, just read the book from start to finish!

To the complete novice, the low-tech webcam approach may well sound crazy. Or you may think I am just advocating astronomy on the cheap, for those with the tightest budgets. After all, there are plenty of expensive astronomical CCD cameras for sale: surely these are better? Well, from a mountaintop observatory enjoying perfectly stable air this might possibly be the case, but, in all other instances, the webcam rules supreme. In the century before digital imaging came along, the visual planetary observer with a quality telescope could see far, far more than the photographer could ever capture, even with huge professional telescopes and the sensitive emulsions of the 1980s and 90s. The reason for this was the Earth's turbulent atmosphere. When you look through a telescope at the Moon or planets they are invariably shimmering and distorting as if submerged in a bowl of water. This is the result of the light from the planet having to pass through the Earth's atmosphere. What a tragedy! For millions or hundreds of millions of miles the light from these planets has hurtled toward the Earth in good shape and then, in the last 30 kilometers, it is distorted by the swirling, turbulent air of our planet. It is like having a 10-meter-deep column of swirling water on top of your pristine telescope mirror! From a time perspective, the light has traveled from distant Saturn for more than one hour to get to the Earth and yet it is distorted only in the last 100 microseconds. Look through any decent-sized telescope at high magnification on a typical night and you can instantly see the advantage webcams possess. The atmosphere blurs, distorts, and ripples the planetary image, but, now and again, there is a fleeting moment of calm, when the atmosphere lets the light pass through to your telescope with little distortion. The eye can spot these good moments because it is imaging all of the time. When that good moment occurs, the observer spots it and the skilled planetary artist makes notes and draws a sketch, based on the best glimpses over many minutes of observing. That is how the keen-sighted planetary observer, with patience and a flair for drawing, worked for centuries. In the last century, a few photographs, or even a few CCD images, would rarely freeze the good moments. Indeed, a few snapshots will not let you even focus a normal CCD camera. Is the image blurred because it is out of focus or because the atmosphere is blurring it? It's impossible to tell. With a webcam like the Philips ToUcam Pro (see Figure 2.1) you can focus in real time, just like looking visually through an eyepiece.

Webcams may be inexpensive, but they have a huge technological advantage. The download speed is blisteringly fast. Even with the original USB 1 standard, the cheapest webcams can transfer 30 frames per second to your PC. This is more than enough to focus with and more than enough to outperform the human eye. However, even a webcam has limitations and we will examine these signal-to-noise considerations in more detail later. If you are still not convinced that a humble webcam, linked to a telescope, can achieve better results than any other detector, then just look at the pictures in this book. As a start, look at Figures 2.2 and 2.3. These are raw and processed webcam stacks taken by planetary imaging master Damian Peach with a 280-mm aperture Celestron telescope costing under $2,000. The final image could almost be mistaken for a Hubble space telescope

Figure 2.1. The Philips ToUcam Pro II, model PCVC 840K.

Figure 2.2. A raw stack of 600 webcam frames of Jupiter, taken by Damian Peach, with a Celestron 11 in Tenerife, Canary Islands.

Figure 2.3. The same image shown in Figure 2.2 has been expertly processed by Damian Peach, using techniques explained later in this book.

image, a telescope costing a billion dollars! The way I look at it is as follows: the webcam is the closest thing that technology has produced to the human eye. It captures images fast and is very light sensitive. Only years of research and development by CCD manufacturers like Sony and electronics manufacturers like Philips could produce such a superb, lightweight, quantum efficient, and inexpensive detector like the Philips ToUcam Pro webcam. In fact, the accumulated research has cost multi-millions of dollars, despite your own webcam costing $100 or less. The only downside to a webcam, compared to a custom CCD camera, is that the CCD chip is not cooled, so long exposures (even if the hardware and software allowed it) would be horribly noisy. However, the Moon and planets are bright objects and long exposures are not needed. We want short exposures to freeze the turbulence of our atmosphere and, in this regard, the webcam rules supreme. Suddenly, we can all get good pictures of the planets, even if we have no artistic ability and lousy eyesight!

Quick Start Guide

This Quick Start Guide is designed for those people who have limited knowledge of webcams or image processing but are comfortable with PCs and gadgets, have a telescope, and just want to get up to speed quick. In other words, they want to get some good lunar or planetary images in the next few days, without studying every chapter of this book. If you fall into this category, this section is for you.

Assuming you have a USB-equipped PC newer than 1998, you need to have the following, all of which can be ordered via the web:

1. A webcam, such as the Philips ToUcam Pro, Logitech QC Pro 3000/4000, or the Celestron NexImage. The supplied software can be used to save webcam AVI videos. PC accessory suppliers sell webcams, as does Amazon. Celestron dealers sell the NexImage.

2. A telescope to webcam adaptor (search the web). The NexImage comes with one.

3. Registax software from http://aberrator.astronomy.net/registax/. Again, the NexImage comes supplied with this, and basic instructions for using Registax are supplied with Registax' help files.

4. A Barlow or Powermate lens, preferably from Tele Vue, to enlarge your image scale. Aim at increasing your focal length to five meters to start with. So, if your basic telescope focal length is 1 meter, use a 5× Powermate. If it is two meters, try a 2.5× Barlow, etc.

5. Once you have all this equipment assembled, you simply install the software as directed (webcam and Registax), carry out some trials indoors (with the webcam lens in place) then connect the webcam and adaptor to telescope and start experimenting. A laptop is useful, as is a separate telescope to find the object. I would also strongly advice experimenting on the Moon first.

6. For more advanced techniques, finish reading this book!

CHAPTER THREE

High-Resolution Essentials

It is not unknown for beginners in this hobby to stare in disbelief at the lunar and planetary pictures secured by the world's best webcam amateur astronomers and lose the will to live. "He must have extraordinary seeing conditions" or "My telescope must be a lemon" are phrases often heard in this context. The factors preventing a planetary observer from achieving the best results are many and varied but, in essence, they boil down to dedication and the following factors (in rough order of importance):

- The observer must do a huge amount of observing to "catch" moments of good seeing.
- The telescope must have precisely collimated quality optics.
- The telescope and observatory should cool quickly to the air temperature.
- The webcam must be focused very precisely.
- Daily information on the stability of the atmosphere/jet stream is vital.
- Image processing knowledge and experience are invaluable.
- No stage of the process should be rushed, especially the image processing.

You will see that I have not specified any particular type of telescope for planetary imaging despite the endless debate on which telescope is best for planetary use. To be honest, the type of telescope is largely irrelevant when compared to the quality of the optics and how accurately they are collimated. An uncollimated telescope is, automatically, a lemon, however good the optics. We will read more about collimation later. The world's greatest planetary observers use a variety of telescope types and get good images with every design. In the main, they use telescopes between 20 and 40 cm aperture, with the very best webcam imagers tending to have instruments in the 23 to 28 cm aperture range (corresponding to theoretical resolutions of around 0.5 or 0.4 arc-seconds). However, some telescopes are definitely better than others from a collimation and user-friendliness aspect. I would place user-friendliness way up there on the list of priorities. Becoming the master of your particular instrument is the key to getting good results. Telescopes are rarely 100% hassle-free unless they are handmade to your requirements. I will have more to say on telescope choices, disasters, and lemons in a short while, but first, I would like to address the factors that I have listed above.

Loads of Observing

I have to be blunt here. Unless you live at a high-altitude professional observatory you are going to have to do a lot of observing to catch the planets at their best. However, if your telescope and observatory are user-friendly and you have a good knowledge of the weather that creates stable air, you will have a distinct advantage. The secret to getting good planetary images is to have an excellent telescope, the use of which you have mastered, poised and ready for those nights of good, stable seeing. But you will still need to go out every hopeful night and check the conditions. The world's best planetary imagers are active every week that a planet is visible high in the night sky. My good friend Damian Peach (a living legend in the field of planetary imaging) often observed on every day of the month when he lived in Tenerife. He freely admits that he used to pray for cloud; but he just could not ignore a clear sky. Such amateurs do not make planet observing a once-a-month hobby. They want to be there when "it" happens. By "it," I mean what another friend of mine (Dave Tyler) calls an "Atmo-calm," when everything is steady and planets look like exquisite paintings seen through the eyepiece. It would not be too extreme to say that planetary imaging needs to become an OCD (obsessive compulsive disorder) to really allow the observer to compete with the world's best imagers. It has to become a nightly ritual to check the weather forecast and "nip out" to check the seeing. Of course, the practicalities of a stressful day job and modern family life may well make this impossible for many. However, if the observer's equipment is friendly enough, a compromise situation can often be reached. In this respect, my fifth bullet point is important. A precise knowledge of the weather and how it affects planetary observing can save a lot of wasted effort.

Before we look at predicting the best planet observing weather though, there is another aspect to "observing a lot." It applies to each night, not just to the number of nights. The best planetary imagers bide their time when they set their telescopes up. Experience has told them that the atmosphere is a moody beast. Even on average nights, if you hang around for hours, drifting in and out of the house, killing time, periodically de-dewing the optics with a hairdryer, there will be a period of a few minutes, now and again, when seeing becomes really good, for no apparent reason. Patience is most definitely a virtue in planetary imaging!

Because atmospheric seeing is such a problem for planetary observers, astronomers have developed seeing scales to be taken into consideration when making observing reports. These scales enable other observers to understand how good or bad the stability of the planetary image was. Of course, assessing the seeing and attempting to accurately categorize how good or bad it is can be fraught with problems. Seeing can vary from minute to minute and comes in different types. Sometimes there is just a slow rippling or distorting effect, othertimes there is little distortion but virtually no detail. The former effects can be due to lower atmosphere effects (or instrumental cooling problems) and the latter are often related to the speed of the upper atmosphere "jet-stream" winds. No seeing scale can ever do more than give a vague indication of conditions. The two seeing scales most used by amateur astronomers are those devised by E.M. Antoniadi (1870–1944) and W.H. Pickering (1858–1938), two legendary planetary observers from the visual era.

Following is the Antoniadi scale, whose numbers becomes larger as seeing deteriorates:

I Perfect seeing without a quiver

II Slight quivering of the image with moments of calm lasting several seconds

III Moderate seeing with larger air tremors that blur the image

IV Poor seeing; constant troublesome undulations of the image

V Very bad seeing, hardly stable enough to allow a rough sketch to be made

Detailed below is the 10-point Pickering scale, whose numbers become larger as seeing improves. It was designed for use with a 5-inch (12.7-cm) refractor. This latter point is important as the Pickering scale describes the appearance of a stellar diffraction pattern and such patterns are far more difficult to see with larger instruments, as the larger the telescope, the smaller the diffraction pattern features are, in arc-seconds. So amateurs applying the Pickering scale with, say, a 25-cm telescope will rarely, if ever, achieve Pickering 10. (Pickering actually invented more than one seeing scale, but the 10-point scale is his best known.)

1. Star image is usually about twice the diameter of the third diffraction ring if the ring could be seen; star image 13 arc-seconds in diameter.
2. Image occasionally twice the diameter of the third ring (13").
3. Image about the same diameter as the third ring (6.7"), and brighter at the center.
4. The central Airy diffraction disc often visible; arcs of diffraction rings sometimes seen on brighter stars.
5. Airy disc always visible; arcs frequently seen on brighter stars.
6. Airy disc always visible; short arcs frequently seen.
7. Disk sometimes sharply defined; diffraction rings seen as long arcs or complete circles.
8. Disk always sharply defined; rings seen as long arcs or complete circles.
9. The inner diffraction ring is stationary. Outer rings momentarily stationary
10. The complete diffraction pattern is stationary.

What are these features called the Airy disc and the diffraction rings? Well, we will learn more about these when we deal with collimation later in this chapter. Essentially, a star never appears as a point source when viewed at high magnification. It appears as a finite disc (the Airy disc) surrounded by a series of increasingly faint rings. The star looks a bit like the splash and ripples resulting from throwing a stone into water (see Figure 3.8). Using a refractor (or a small obstruction reflector), on nights of perfect seeing, a textbook-perfect Airy disc and rings will be seen. These diffraction effects limit the resolution of an optical telescope such that a 12-cm aperture might resolve roughly 1 arc-second, and a 24-cm aperture might resolve 0.5 arc-seconds, or so. Pickering was describing the distortions to the perfect diffraction pattern that he saw when the Earth's atmosphere was turbulent, and he allotted the scale above to what he saw through his 12.7-cm refractor. We will learn more about resolution theory in Chapter 7, when we prepare to attach our webcam for the first time.

Predicting The Atmo-Calms

The popular concept of an astronomer's perfect night is one where a cold front has swept through the observer's location and a crystal-clear sky full of twinkling stars can be seen. In fact, this is a nightmare scenario for the planetary observer. A cold front passing through a region may well reduce the moisture content of the air and is great for looking at deep sky objects and comets, but it leaves the air in a very unstable state and, invariably, the hotter ground radiates its daytime heat into space, further increasing the chaos in the atmosphere. At high powers, the Moon and planets will be a wobbling, distorting mess. What is required for the best planetary views is stability, and this invariably comes from the presence of a high-pressure system anchored over the region. In a high-pressure system, especially one that has been around for a few days, the air becomes hazy (and polluted in cities) but it is very stable. The ultimate in atmospheric stability is often achieved when mist or fog is forecast. Of course, this invariably leads to a heavy dew settling on every surface and, when mist turns to fog, the planets become too faint to image. However, just before the mist turns to fog, exquisite planetary views can be seen. As a keen planetary imager I get pretty excited when there is a high-pressure system over the U.K. and fog is forecast. Time and time again, my best views have been acquired just before the planet faded away as the fog thickened. It is not surprising that this should be the case. Fog is the ultimate proof of atmospheric stability. A high-pressure system is not essential for planetary imaging, but low wind speeds at all levels in the atmosphere seem to be crucial. This is the experience of almost every planetary observer. Of course, sitting in the middle of a high-pressure system reduces wind speeds to zero, but so does being in a "Col," that is, sitting between two high-pressure and two low-pressure systems where little is changing and winds are low.

The Earth's lower atmosphere is only half the problem, however. There is another issue that does not appear on national TV weather forecasts. This is the issue of the Earth's upper atmosphere jet stream. The jet stream is, without a shadow of a doubt, the "fine planetary detail" wrecker. It might seem incredible that an area of the Earth's atmosphere that is 8 to 10 kilometers above the surface and where the pressure is only 300 millibars could possibly affect our view of planets so badly. However, the light from the planets has to pass through the jet stream altitudes and, in extreme cases, the jet stream wind speeds can be as high as 500 kilometers per hour. When high jet stream winds are over your observing site, they will appear, in a defocused planetary image, like a river of water streaming through the eyepiece. This will result in extremely "fast" seeing, with no hope of even a webcam freezing the planetary details. Under such circumstances, my advice is to go to bed. Nothing useful can be achieved!

Fortunately, via the Internet, various companies publish jet stream weather forecasts, principally for aviators. Many of these forecasts originate from data compiled at the U.S. National Center for Atmospheric Prediction (NCAP), and predictions are issued twice daily for sites all around the world. Any Internet search engine should be able to find a site predicting jet stream activity for your location without too much trouble. Such forecasts can save the planetary observer a lot of wasted effort. The Unisys Aviation pages (currently located at http://weather.unisys.com/aviation) are the best pages currently on the web for studying the wind speeds at various atmospheric heights all over the Earth's surface (see Figure 3.1). The 300-mb pressure pages are the ones that the planetary imager should study in detail, and that altitude is not covered by TV weather forecasters. Obviously, one is looking for the jet stream equivalent of a lower atmosphere high pressure system centered on the observer's site (i.e., low wind speeds). Quite often this will occur when a standard high-pressure system has been in place for a few days, but not always. At high northern temperate latitudes, winter often brings a polar jet stream down from the arctic, even when a high-pressure system is not far away. The jet stream forecasts, compiled for the aviation industry, are the nearest thing that amateur astronomers have to a nightly "seeing" forecast. High jet stream winds can wreck seeing even when a high-pressure system is nearby. However, after a sunny day, turbulence can be poor regardless of other factors, as the hot ground can radiate rapidly. There is considerable evidence that a stable state forms in all levels of the atmosphere (even in the jet stream) immediately after sunset. There appears to be a delay before temperatures plummet, in which good planetary images can be secured. Often, this means finding the planet in bright twilight, only half an hour after sunset. Another calm period exists in dawn twilight, too, when the nighttime cooling has slowed down to a minimum.

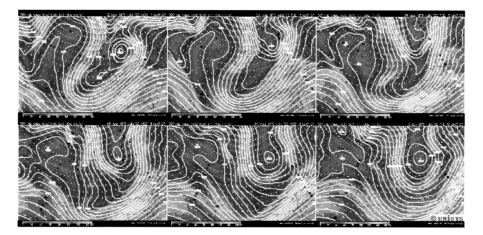

Figure 3.1. Jet stream weather maps. Purple = low winds; yellow = high winds. Image: Unisys.

Apart from the jet stream sites, two other web addresses may be of interest. Firstly, a huge amount of relevant data regarding wind speeds from sea level to a height of 9 kilometers is available at Meteoblu at http://pages.unibas.ch/geo/mcr/3d/meteo/index.htm. There is also a weather chart archive, where you can retrospectively check the conditions on a given night at http://www.meteoliguria.it/archivio21.asp.

Damian Peach cross-checked his best U.K. seeing nights against this page and came up with the following table:

Date	Seeing	Sea-Level Pressure	300-mb Wind Speeds
1999/10/06	Pickering 8-9	High 1028 hPa	12 m/s
2000/10/13	Pickering 8-9	Low 1002 hPa	12 m/s
2003/09/29	Pickering 8-9	High 1016 hPa	20 m/s
2003/12/16	Pickering 8-9	High 1030 hPa	18 m/s
2004/03/01	Pickering 8-9	High 1036 hPa	20 m/s
2004/04/14	Pickering 8-9	High 1018 hPa	12 m/s
2004/10/01	Pickering 8-9	High 1020 hPa	17 m/s
2004/12/11	Pickering 8-9	High 1026 hPa	12 m/s
2005/01/13	Pickering 8-9	High 1032 hPa	14 m/s

All of these nights featured low wind speeds at sea level and 300-mb jet stream wind speeds of 20 meters per second or less. The September 29, 2003, "event" was one I well remember. The U.K. was under a "Col," not, strictly, a high-pressure system, just a region of inactivity between high- and low-pressure systems.

Locations

Amateur astronomers often wonder whether there is any ideal place to live to secure good stable conditions, and I will have more to say about this later in chapter 5, "Have Webcam, Will Travel." Obviously, a high-altitude site near the equator is ideal, as the planets will then pass almost overhead and the light will pass through relatively little air. The light from a high-altitude object is also considerably less dispersed (split into colors) than from an object at, say, 20 or 30 degrees altitude. However, most amateurs simply do not have the luxury of moving abroad. Living on the slopes of a mountain or a hillside can have good and bad effects. If you go for this option you do not want to be sited on the leeward side of the hill. If the prevailing winds come from the west and you are on the east slope of a hillside, turbulent air will cascade past your observatory, wrecking the seeing. Coastal or island sites can be excellent for planetary observers, especially when the prevailing air flow is off the sea. The sea varies much less in temperature than the land and is virtually flat. A light laminar air flow off the sea can produce excellent coastal seeing, although with the risk of nighttime sea fog. Areas prone to atmospheric inversion (i.e., the temperature in the atmosphere increases, rather than decreases, through a limited layer of air) are often excellent for stable planetary observing, too. In his 1995 book *High-Resolution Astrophotography*, Jean Dragesco mentions peculiar atmospheric inversion conditions in Zaire that occasionally lead to temperature variations as small as 1.5°C from sea level up to the jet stream. A planetary imager's dream!

Collimation

I never fail to be astounded at the number of amateurs who do not keep their telescopes collimated. I wonder whether they are simply scared of doing more harm than good, or even damaging the instrument. Admittedly, the traditional methods of star collimation are best done with two people, but, for the webcam user, this is just not necessary. It must be said that not all telescopes, even expensive ones, have friendly mirror adjustment systems. I know of one world-class planetary imager who bought what he thought was the ultimate planetary telescope (at a cost of $13,000 for the optical tube assembly) only to find that the mirror collimation system was so complex that the telescope was virtually unusable. This was a telescope with excellent optics, made useless by the dealer's bodged primary and secondary mirror cells. The telescope was so bad in this regard that when the tube was horizontal the primary mirror was prone to being catapulted out of its cell by the spring-loaded mirror support! However good a telescope manufacturer is at making optics, make sure they or their dealers can make reliable adjustable mirror cells too.

Precise collimation is a necessary chore because even the finest telescope only delivers pin-point diffraction limited images over a very narrow field of view. This will be irrelevant for low-power views of galaxies and comets, but absolutely critical for high-power views. In this regard, the finest planetary telescopes are the long-focus Newtonians. I am talking here about instruments with (typically) apertures of 25 cm or so and focal ratios of 7 to 10 (i.e., focal lengths between 1.75 and 2.5 meters). A term you will often hear about in this context is the "sweet spot." The sweet spot of a planetary telescope is the diameter of the region within which perfect pinpoint star images appear. As you move further away from the optical axis of the telescope, stars become distorted by aberrations like coma and astigmatism. As you cross the sweet spot boundary, these aberrations just start to perceptibly degrade the star images. Needless to say, your planet needs to be within the telescope's sweet spot, ideally, bang in the middle. The physical size of this sweet spot is frighteningly small. With a simple Newtonian, coma starts to degrade the image first and the sweet spot diameter is proportional to the cube of the telescope's f-ratio. (Yes, you did read that right—the cube!) Just to give a few examples, an f/4 Newtonian (of any aperture) will only have a sweet spot that is a tiny 1.4 mm in diameter! An f/6 Newtonian will have a 4.7-mm sweet spot. Increase to f/8 and you get a very nice 11.2-mm sweet spot. If you are a real long-focus Newtonian fanatic, f/10 will net you a whopping 21.9-mm sweet spot. Converting these diameters to angles, for a 250 mm aperture, gives you sweet spot angles on the sky of 4.8, 10.8, 19.3, and 26.5 arc-minutes. As can be seen, the 250-mm f/4 Newtonian has a sweet spot barely larger than the largest lunar craters. Conversely, the 250-mm f/10 Newtonian has a sweet spot almost as large as the full Moon!

The Newtonian is only one type of telescope. Compound telescopes (like Schmidt-Cassegrains, Maksutovs, and Maksutov-Newtonians) have a variety of sweet spot diameters, but they are all measured in arc-minutes and they all need collimating precisely for good planetary performance. The ever-popular Schmidt-Cassegrain Telescope (SCT) design can only be collimated by adjusting the tilt of the secondary mirror to reflect the optical axis of the primary straight down the drawtube middle; adjusting the primary mirror in such telescopes is not practically possible.

The final stage in the collimation of any planetary telescope has to be done on either a real star or an artificial star. This can be frustrating at first because, as you adjust the optics, the star will move, so it needs to be recentered. However, after a bit of experience, the collimating chore can become routine. With fine adjustments the test star will only move an arc-minute or two. Of course, if manufacturers made telescopes whose optics did not move around, you would only have to collimate a telescope once. Sadly, this happy state of affairs rarely exists unless you build yourself a custom long-focus Newtonian. Schmidt-Cassegrain telescopes have "conical" primary mirrors (i.e., they are thinner at the edges than in the middle). Such mirrors' relatively light weight can be supported by the telescope's central baffle tube alone. However, this raises the problem of SCT mirror flop. As you move the telescope around the sky, or flip the telescope about the declination axis of your German Equatorial Mounting (GEM), the mirror position will change. It may only change by a few arc-minutes, but that is enough to wreck the collimation. Fortunately, the three collimation screws on an SCT's secondary are easy to adjust, even if you do have to remove the dew cap to get at them. I know of one planetary imager who always leaves his SCT parked in the position he images in (i.e., on the meridian, at the planet in question's declination, so he can retain collimation).

The modern webcam imager has a distinct collimating advantage over his visual counterpart, especially when using a telescope that is pretty close to perfect collimation to start with. You can simply attach a webcam to a high power Barlow lens or eyepiece and observe the test star on your computer screen, as you tweak the collimation for that final stage. You can have the slow-motion hand controller next to your PC to keep recentering the star too. This makes life a lot easier than darting from eyepiece to adjustment screw, back-and-forth, back-and-forth, bending and stretching while sweating away in thermal clothing! With a webcam and a laptop, everything can be done in relative comfort.

Let us now examine, in detail, the precise process of collimating a telescope from scratch and the tools that help to make this process relatively painless.

Basic Collimation of a Newtonian Reflector

Figure 3.2 shows the three phases in the basic daytime collimation of a Newtonian telescope: uncollimated, secondary mirror adjusted, and primary mirror adjusted. The Newtonian is an ideal telescope to collimate precisely because both primary and secondary mirrors can be adjusted; everything can be made "just right." The figure shows the view, through the drawtube, when the observer's eye is centered where the eyepiece normally sits. A plastic 35-mm film canister with a hole in the middle makes an excellent sighting tube for positioning the eye. Just follow the steps in the figure and the basic collimation is complete.

The best way to visualize the optical configuration of a Newtonian telescope, from a collimation viewpoint, is that there are two optical axes: the optical axis of the primary mirror and the optical axis of the eyepiece. The axis of the primary mirror is at right angles to the primary at the optical center of the primary (this center is usually assumed to be at the dead center of the circular glass mirror). If you have a thoughtful mirror manufacturer they will have marked this center with

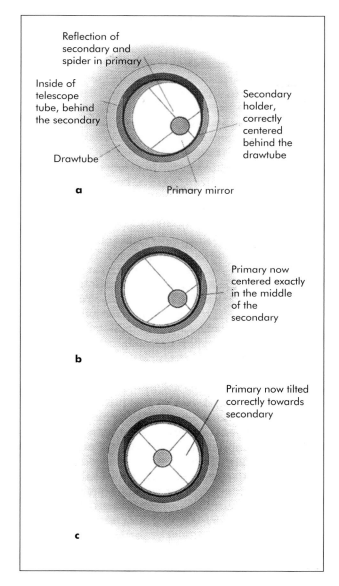

Figure 3.2. The three key stages in rough collimating a Newtonian reflector, prior to star collimation – this is far less critical in a slow Newtonian.

a tiny dot or ring. If not, it will help to mark the mirror yourself. As the dead center of the mirror lies under the "shadow" of the secondary mirror, this marker will *not* affect the performance. The photons from a star in the precise direction of the primary mirror's axis will be reflected and "focused" to a perfectly sharp image at the focus on this axis. (This is where your webcam needs to ultimately sit.)

The axis of the eyepiece is usually assumed to be at the center of the focuser drawtube. Of course, for the webcam user, it will really be the dead center of the webcam chip, after the obligatory Barlow or Powermate enlarging lens. But, when we are actually collimating the system before the star test, the eyepiece axis will, effectively, be the axis of the observer's sight tube, Cheshire eyepiece (read on), or laser collimator (also read on). The secondary mirror diverts the incoming light to the side of the Newtonian tube. The secondary will also divert the optical axis of the primary and the eyepiece (depending on how you look at the situation).

The main purpose of collimating is to align the primary and secondary mirrors to form one common axis. Normally, you do this by adjusting the position and tilt of the secondary mirror, and the tilt of the main mirror. In addition, the drawtube should, ideally, be parallel to the primary mirror's optical axis. With webcams, one has to assume that the chip inside the webcam is mounted at right angles to the webcam-drawtube adapter. Let us not get too paranoid at this stage!

One issue that often worries first-time Newtonian collimators is that when they look down their telescope tube they get the impression that the secondary mirror is not exactly centred in the tube; the flat secondary mirror appears to be offset toward the mirror end and away from the eyepiece end. In fact, this is perfectly normal, as the light cone that needs to be captured is slightly bigger at the bottom of the flat secondary mirror than at the top. This "offsetting" of the secondary mirror is routine in quality telescopes but the required offset is fairly small in long focal length Newtonians where the size of the secondary is so much smaller than the focal length. Some telescopes have offset secondary mirrors (fast, f/4 Newtonians and telescopes with tiny secondary mirrors really need them), some do not. The critical point here is that, when looking through a sighting tube at the reflection of the primary mirror in the secondary mirror, you should be able to see the whole of the primary. If you cannot, light is being lost. Ideally, the reflection of the primary should appear concentric in the secondary mirror. The offset formula is: minor axis/(4 × focal ratio). So for a 30-mm minor axis secondary on an f/7 Newtonian you need to slide the flat mirror 30/(4 × 7) = 1.07mm toward the primary and away from the eyepiece. In this case, you would need to be a perfectionist.

The good news for the planetary webcam imager, who is only interested in capturing objects in a field that is only 1 or 2 arc-minutes wide, is that once you have carried out the two-stage mirror alignment shown in Figure 3.2 you can move straight to collimating on a star (or even an artificial star). Collimation purists may well disagree with me here, but I am trying to steer a course between not collimating at all (seemingly, the approach of 90% of amateurs) and obsessive collimation. Yes, you can ensure that the drawtube/focuser is perpendicular to the tube (by removing the secondary holder and passing a pipe across the tube or by using a laser to project across the telescope tube). Yes, you can redesign the secondary support so that the secondary is offset precisely. But, to get those perfect planetary images you simply need to: 1) carefully tilt and rotate the secondary mirror until the primary mirror appears concentric within the secondary; 2) tilt the primary mirror until the secondary mirror is concentric in the primary; 3) conduct a star test. In stage 2, it is quite possible to achieve perfect collimation with a long-focus Newtonian by sheer good luck. This is because, if you have a spot marked on the mirror's optical center and can see the reflection of your eye and the sight-tube hole centered on that mirror spot, you may well be within a sweet spot that is 6 or 7 mm across. However, this cannot be relied upon. A star test *must* be performed, ideally using the equipment with which you will image.

Collimation Aids

Before we move onto the star test itself, I would like to say a few words about "gadgets" to help you collimate a telescope. The cheapest gadget, and one that can easily be home made, is a basic sighting tube as shown in Figure 3.3A. All this needs to be is a device just 31.5 mm in diameter (to fit the 31.7 mm diameter of a 1.25 inch eyepiece hole) with a small hole placed dead in the center, a few millimeters wide. This device ensures that the observer will hold their eye at the dead center of the drawtube when rough-collimating the secondary and primary mirrors. A useful addition is to place a white ring on the inside of the sighting device, around the observer's eye hole. This white ring will be obvious in daylight and its reflection will aid collimation.

Another gadget, available commercially, is a device known as a Cheshire eyepiece. This is little more than an eyepiece-sized sighting tube with a reflective washer-like insert, tilted at 45 degrees, and inserted into the sighting tube. It looks

Figure 3.3A. The Orion (USA) Collimating eyepiece is, essentially, an accurately machined tube with a central sighting hold mounted within a bright metal disc. Image: Jamie Cooper.

a bit like a miniature half-periscope. A gap in the side of the Cheshire allows light to be directed onto the reflective washer, creating a bright, doughnut-like reflection on top of all the other reflections seen during the collimation process. If the primary mirror center is marked with a spot or a ring, it enables accurate alignment of the primary spot with the doughnut's reflection, thus aligning the optical axes.

A more modern gadget is a laser collimator. Again, shaped in an eyepiece-sized package, this device shines the beam from a laser diode in a pencil-thin beam from the drawtube, onto the secondary and then onto the primary. If everything is lined up correctly, the laser beam will trace a return path exactly along the path of the beam, ending up with the laser beam pointing back at itself. With some laser collimators, the point where the return path strikes the laser is made easier to see by combining it with a Cheshire type design (i.e., a hole in the laser body side through which the laser's return journey can be verified). Such a device is shown in Figure 3.3B. Some caution is needed with laser collimators though, and not just from the point of view of eye safety. Unlike placing your swivelling eye at the hole of a sighting tube, the laser is simply blindly shooting its beam straight out of its casing. It is therefore wise to check that the telescope's drawtube really is perpendicular to the telescope tube and that the laser beam is emitted parallel to the laser beam collimator body. Some years ago, *Sky & Telescope* magazine tested a laser collimator whose beam was *not* emitted parallel to the collimator body. This would lead to an erroneous collimation. To check the machining accuracy of the laser collimator, mount the laser collimator in the telescope drawtube *backward* and point the laser at a wall (with the telescope drive off). Then rotate the collimator in the eyepiece. The spot on the wall should *not* move as the collimator is rotated. If it does, your collimator is a

Figure 3.3B. The BC&F Astro-Engineering Laser Collimator combines a laser collimator with a "Cheshire" type sighting hole, so you can see exactly where the laser beam is being directed. Image: Jamie Cooper.

lemon, unless it has some provision for adjusting the beam. This test is easier to perform if a spare drawtube is mounted rigidly in a workshop bench vice. After verifying (hopefully) that your laser collimator is OK, removing the secondary mirror and seeing where the laser beam ends up on the opposite side of the telescope tube will verify whether your drawtube and focuser is square-on to the tube. Personally speaking, I have never been that attracted to laser collimators. I have always reckoned that visually collimating by eye and then precisely collimating on a star is the best sequence of events. Laser collimators are often hyped as the ultimate collimating tool, but if their limitations are not understood they can be of limited benefit (don't believe all of the manufacturer's hype). Unfortunately, collimating on a star can be thwarted by poor seeing, which is why I like the next gadget so much.

BC&F Astro-Engineering's Picostar, shown in Figure 3.3C is an artificial star generator that produces a point source illumination out of a 50-micron diameter fiber-optic ferrule. It allows a variety of illumination levels and, provided you have a long garden and a telescope that stays in collimation when the telescope is moved around, it is a godsend. For the device to work properly, it needs to be far enough away from the telescope that the 50-micron aperture appears smaller than the diffraction limit of the telescope. In practice, this means that a 200-mm instrument requires the Picostar to be placed at least 20 meters away and a 300-mm instrument needs the *Picostar* to be 30 meters away. In practice, many telescopes will need the device to be that far away anyway, as otherwise they could not achieve focus. With a Newtonian you may well need a drawtube extension tube to achieve focus. I have heard this type of adaptor referred to as a pervert tube as it enables a Newtonian to be focused on bedroom windows. It will not surprise anyone to learn that I own one.

In the pages above I have discussed, at length, the basic daytime collimation of a Newtonian reflector. This is because in home-made or commercial Newtonians, all of the component parts are often able to move and are all accessible for adjustment.

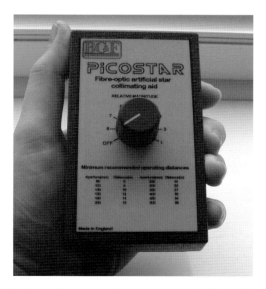

Figure 3.3C. The BC&F Astro-Engineering Picostar. An invaluable artificial star device with a 60-micron diameter light source. Image: Martin Mobberley.

The Newtonian is almost unique in this respect. With other telescopes, especially mass-produced ones, things are rather different. Refractors are largely collimated for life. The relatively light and long focus optics of commercial refractors are often fixed and the optics of mass-produced small Maksutovs are frequently fixed too (although in some cases they would be much better if they were adjustable). The ubiquitous Schmidt-Cassegrain, whether made by Meade or Celestron, has no provision for adjusting the primary. As we have already seen, the SCT primaries will often tilt by an arc-minute or two as the telescope is moved and thus nightly collimation becomes necessary. The optical axis of the primary mirror of an uncollimated commercial Schmidt-Cassegrain usually ends up pointing within a few millimetres of the centre of the instruments secondary, and a few minutes tweaking on that mirror's adjustment screws, will produce perfect collimation, at least, until the primary flops around a bit on it's central support. With a Schmidt-Cassegrain, there is only one daytime collimating step, adjusting the three screws on the secondary mirror (Figure 3.4) until the reflection of the primary in the secondary is concentric, as seen through a drawtube sighting tube. With regard to SCT mirrors "flopping" out of collimation, the larger the SCT, the

Figure 3.4. The collimation screws on a Schmidt-Cassegrain can be found on the secondary holder mounted on the corrector plate. Image: Martin Mobberley.

more likely this is to happen, in my experience. I would like to say one thing in favor of SCT collimation though. Most Newtonian and Cassegrain systems have push-pull collimation systems. In other words, you fiddle about with two thumb-wheels to get collimation: you slacken one thumbwheel then tighten the other at the primary mirror end. This can be a real nighttime hassle. Although you can only adjust the secondary with an SCT, the secondary mirror is light and the adjust-ment screws are spring-loaded. Therefore, you just tweak one of the three screws (or, ideally, small thumbwheel/knob screws) and the mirror moves. There is none of this push one way/pull the other hassle at the mirror end. Telescope makers please note! SCTs are a breeze to recollimate, even if they rarely stay collimated.

Nighttime Star Collimation

After daytime collimation, the next step for any telescope is nighttime star collima-tion. If you have a rigidly made Newtonian, a long garden, and an artificial star device, you need not bother about real stars or a clear night. An artificial star has a huge advantage over a real one—atmospheric seeing is not an issue. The textbook diffrac-tion patterns will always be seen, even if tube currents and heat from the ground mean they are not quite perfect. For both Newtonian and Schmidt-Cassegrain colli-mation, the old-fashioned visual way of collimating on a star is a hundred times eas-ier if an assistant is employed, so the telescope owner can look through the collimating eyepiece while the assistant adjusts the mirrors. However, as I have already mentioned, collimating with a webcam and a high-power Barlow is a much easier system. But the choice is yours. Either way, you will need to know what to see.

The first stage in star collimating a reflecting system is probably unnecessary if careful daytime calibration has already been carried out. It simply involves observ-ing a first magnitude star at a decent image scale (say 250× visually), and well out of focus, and checking that the black hole (the shadow of the secondary mirror) is in the middle of the star's out-of-focus disk. Figure 3.5 shows the real image of an out-of-focus star, imaged with a well-collimated Newtonian, by expert imager Mike Brown. Note that the real image looks a bit different to the perfect simulation in Figure 3.6. That figure, like most of the collimation images here, was produced using a software simulation package called Aberrator (http://aberrator.astronomy.net/).

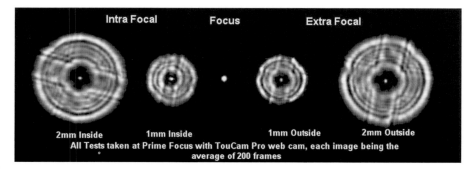

Figure 3.5. A real star test on a quality 200-mm Newtonian primary mirror made by Orion Optics (U.K.). The star is examined inside (intra) and outside (extra) focus. Image: Mike Brown.

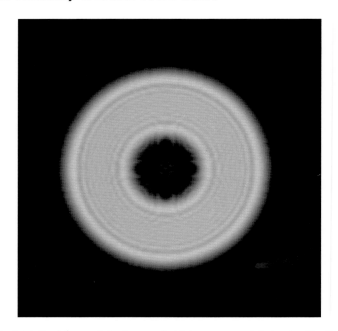

Figure 3.6. A totally defocused star in a perfect telescope, simulated using Cor Berrevoet's Aberrator software.

If the black hole is not in the middle you are way off collimation and the mirror collimation screws need to be adjusted until the shadow is centered.

The second stage in star collimation is far more demanding and can only be carried out when seeing conditions are reasonable (or with an artificial star). Typically, a second or third magnitude star is chosen (for a 200–300-mm aperture) and it must be well above the horizon so turbulence is minimized. A very high magnification is then used (600× or more for the visual observer and maybe 0.1 arc-seconds per pixel with a webcam) and the star is moved from well inside to well outside of focus while the diffraction rings are examined. There should be a bright dot in the middle and then a series of concentric dark and light rings out from the center. As the scope is moved through focus, this pattern should open and close smoothly and symmetrically (Figure 3.7 shows the slightly defocused view). If it fails this test, the mirror adjusting screws need tweaking. Of course, every time the screws are tweaked, the star will move and will need recentering in the field. Once the intra- and extrafocal patterns resemble a perfect textbook diffraction ring, the third step can be carried out.

The third and final step to perfect collimation can only be executed when seeing conditions are near-perfect—a rare event for most people. The set-up is the same as for step 2, except that the star is perfectly focused. We are now looking for the perfect Airy disk, a so-called "false" central disk, surrounded by diffraction rings of diminishing brightness (Figure 3.8). If the first ring is not uniform, or is incomplete (as in Figure 3.9), the collimation screws need tweaking by a tiny amount to achieve a complete and uniform first ring. This last test is so sensitive that even

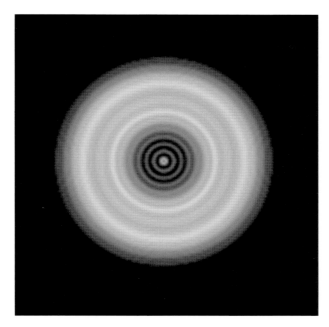

Figure 3.7. A slightly defocused star in a perfect telescope, simulated using Cor Berrevoet's Aberrator software.

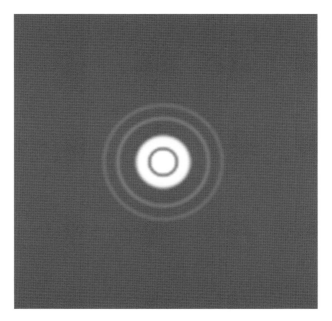

Figure 3.8. A perfectly focused star in a perfect telescope, with perfect collimation! A very high magnification is needed to see this view, along with a night of excellent seeing.

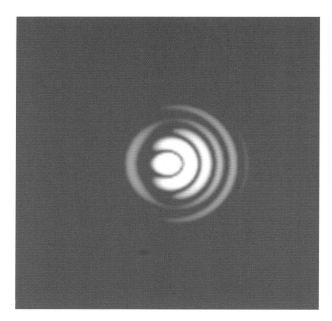

Figure 3.9. A perfectly focussed star in a perfect telescope, but fractionally out of collimation. Note the difference between this and Figure 3.8. A very high magnification is needed to see this view, along with a night of excellent seeing.

moving the scope around the sky will alter the situation with many telescopes! Hard-core planetary imagers will often tweak their telescope's collimation with this final step, before they commence imaging and even adjust the collimation throughout the night.

Well, we have now covered collimation, but it should be emphasized that for the star tests, the diffraction patterns seen will not resemble anything like the textbook patterns unless the night is perfect or you are using an artificial star. If you want to see textbook diffraction patterns on a less-than perfect night, try looking through a small aperture quality refractor or Maksutov at a very bright star. This will enable you to familiarize yourself with what should be seen. Users of giant Newtonians, e.g., 40 cm and larger, will have to accept that they may never see a perfect diffraction pattern or Airy disk unless they stop the instrument down. Not only are the transient, perfect, atmospheric columns (often called "cells") of stable air rarely more than 30-cm across, large telescopes take a very long time to cool down.

Of course, if even on a good night, when stars are not shimmering wildly, the Airy discs look distorted, and you have been through the telescope collimation process carefully, you may deduce that you have inferior optics. To be honest, this is very rare with optics from any of the major manufacturers. Competition is fierce in the modern telescope market and companies just cannot risk turning out poor optics. However, if you are using an old telescope, one from a backstreet mirror "cowboy," or one of dubious secondhand origin, the optics might be suspect. However, in my experience, most amateurs who complain of poor optics simply have failed to collimate them to the precision required to get diffraction-limited views of the planets. Also, a mirror cell that "pinches" the optics is often a source of nonperfect star images.

It can be fascinating to examine a perfect artificial star with a perfectly collimated Newtonian. The act of merely taking the rear mirror cell dust cap off, thereby altering the tube currents, can cause dramatic changes in the diffraction patterns of the star and makes you appreciate just how important the thermal properties of a telescope are.

Thermal Considerations

If someone were to tell you that having an electric fan on a telescope was, perhaps, the most important decision you would make in your observing career, you would probably brand them as insane! But, as the years have gone by (and I have been observing planets for over 30 years) I have become increasingly convinced that the ability of a telescope to cool to the night air is absolutely crucial to getting good results. Back in the 1970s, the French optician Jean Texereau claimed that a temperature difference of 1/7th degree Celsius in a telescope tube could perceptibly alter the performance of the instrument. This would not surprise me. In terms of optics cooling to the night air there are two prime considerations. Firstly, the turbulence created inside the telescope tube, especially near to the mirror; secondly, the distortion of the optical components, as they cool down. Prior to the mid-20th century, most amateur telescope mirrors were made from plate glass with considerable temperature expansion coefficients. It was common for telescope-makers in those days to undercorrect their parabolic Newtonian mirrors slightly (correcting in this context means correcting from a spherical to a parabolic surface) such that the cooling of the plate glass at night would cause contraction to "snap" the mirror into the perfect shape early in the observing session. However, since the introduction of Pyrex mirror blanks, the contraction of telescope mirrors has not been a major problem, unless you are still using a historic instrument with a plate glass mirror. In recent years the (admittedly expensive) option of the glass called Zerodur has meant that amateur telescope mirrors with a zero coefficient of expansion are available. So, when we are talking about the detrimental thermal effects of a warm telescope mirror in the 21st century, we are talking solely about the turbulent air problem and *not* the shape of the mirror.

In 2004 I acquired an excellent 250mm f/6.3 Newtonian reflector from the U.K. manufacturer Orion Optics. It featured a relatively thin (by historic standards), 8:1 ratio primary mirror mounted in a lightweight metal tube. It also featured a cooling fan built into the primary mirror cell (Figure 3.10). The whole optical tube assembly only weighed 11 kilograms. I was astounded (and still am) at the performance of that telescope. I had previously owned 36- and 49-cm reflectors with good optics as well as 30- and 35-cm SCTs and yet this relatively modest aperture Newtonian gave me sharper planetary views than any of those larger telescopes. As far as I can judge, there are three reasons for the lunar and planetary performance of this telescope. Firstly, the optical quality; the primary mirror has a 1/45th wave RMS surface, verified by a Zygo interferometer (I have more to say on this subject in a few pages' time). It also has a Strehl ratio of 0.981. (A Strehl of 1.0 defines a perfect unobstructed primary where 84% of the light from a star goes into the Airy

Figure 3.10. The cooling fan on the rear of the author's 250-mm f/6.3 Orion Optics Newtonian. The mirror is just over three centimeters thick and cools to within a fraction of a degree of the night air in under one hour of fan cooling. Image: Martin Mobberley.

disc and 16% into the rings; 0.981 is very good.) Secondly, the aperture is ideally suited to exploit the best resolution achievable from the U.K. I never feel I need to stop the aperture down and I rarely get multiple images, even in poor seeing. These were common in the 36- and 49-cm reflectors. Thirdly, the telescope has a low thermal mass. I feel that this third point is crucial and, possibly, as important as the other two, because the seeing has always seemed pretty good through this telescope. This indicates that a lot of my seeing problems with larger telescopes were instrumental, not atmospheric.

Telescopes that are larger than about 25 cm in aperture tend to suffer from serious cool-down problems, especially in the evening, when nighttime air temperatures can drop rapidly. A mirror that is more than about 30 mm thick will have serious difficulty adapting to the temperature of the night air, which can drop by several degrees Celsius per hour after sunset. The mass of a telescope mirror increases with the cube of its diameter, assuming its thickness is maintained at the same value as aperture increases. However, the surface area of such mirrors, vital for radiating the heat away, only increases with the square of the diameter. Thus, unless the thickness of a telescope mirror is fixed at, say, 30 mm (necessitating a sophisticated mirror support system for large, thin mirrors, to stop them bending), serious thermal problems will arise in mirrors of 30-cm aperture and above. This problem was investigated in detail in the 1960s by the British mirror-maker Jim Hysom. He found that the mirror core temperature for 30-, 45-, and 76-mm-thick mirrors cooled at typical rates of 3.3°, 1.6°, and 0.9°C per hour after the onset of night. Bearing in mind the air temperature can fall at more than 3°C per hour in the early evening, it can be seen that mirrors more than 30 mm thick cannot even match the falling air temperature with both front and rear faces exposed, unless fan cooling is employed. One solution to this problem is to use conical telescope mirrors, such as used in SCTs. In such mirrors, weighing about 60% of a conventional mirror, the mirror is made from a honeycombed cantilevered structure with the mirror edges thinner than the center. Mirrors up to 40 cm or so in aperture can be supported at the center only with this design and cooling is typically twice as rapid as in a conventional mirror. The renowned U.S. optician William Royce offers mirrors of this type in a Newtonian format.

The most common solution to the mirror and tube current thermal problem is to use lightweight vibrationless fans. In my own 250-mm Newtonian, this fan sits behind the primary mirror. However, a slightly better solution is to employ additional filtered fans in the side of the telescope tube, blowing air across and sucking air out of the region just in front of the primary mirror, where the worst turbulence occurs. Newtonian telescopes are renowned for having tube currents. But if the telescope tube and optics are all at the same temperature these currents tend to disappear. There are two schools of thought on what to do with the problems arising from the telescope tube itself cooling down. One approach (as with my 250-mm instrument) is to have a very thin-walled metal tube (1 mm in my case) that cools to the night temperature rapidly. Another approach is to have the tube manufactured from materials with insulating properties so that they only radiate heat slowly. Historically, mahogany tubes lined internally with cork have been advocated and self-adhesive cork tiles are available from good DIY stores. Regardless of this, a good approach is to have the telescope tube of much larger diameter than the primary mirror and to have a gap in the center of a long tube, to let air in. Completely open tubes are excellent thermally, but terrible from the perspective of dew, dust, and insect ingress. They also allow the observer's body heat to drift across the light path. My ultrathin-walled metal tube seems to be a good compromise and is light, too. I have thermocouples on the tube of my telescope and on the top edge of the primary mirror (Figure 3.11). After some 30 minutes of fan cooling the telescope is at equilibrium and the lunar and planetary views can be exquisite. It took me some 30 years to appreciate how important telescope cool-down was and why mirrors larger than 250 mm have a real thermal problem. Fortunately, the Earth's atmosphere rarely allows resolutions finer than 0.5 arc-seconds to be achieved anyway, so a 250-mm telescope is pretty much all you need in the webcam era. In passing, I would like to add that the renowned Florida planetary observer Maurizio Di Sciullo recently revealed to me that he has to use a 20-cm fan on his 250-mm (40-mm-thick) Newtonian mirror to cool it to ambient temperature. He has concluded that the only way he can ever cool his new 360-mm (60-mm-thick) planetary telescope's mirror to ambient temperature, within the duration of a night, is by water-cooling the primary; fan-cooling is not sufficient! Mirror cooling is a very serious issue.

Different telescopes have different thermal properties. Newtonian telescopes can behave very badly if the tube is sealed at the mirror end. An open mirror cell design, with the glass mirror back open to the air is essential. On my Newtonian, capping the mirror end while examining a star at high power, and then uncapping

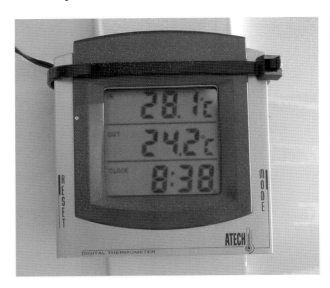

Figure 3.11. A cheap indoor/outdoor thermometer attached to the telescope can be used to measure the mirror temperature (via a lead) as well as the tube or air temperature. These devices are so cheap and light that several can be employed all over the telescope. Image: Martin Mobberley.

it, is dramatic! As soon as a free air flow through the tube end is available the star's diffraction pattern becomes near-perfect on good nights. With the mirror end capped, stars distort into a tear-shaped drop, due to air currents seeking a way to escape. Large Schmidt-Cassegrain and Maksutov telescopes (above 250 mm in aperture) can store considerable amounts of heat in their glass components and in the trapped air inside the instrument. To alleviate this issue, one manufacturer has even designed a probe that can be inserted into the eyepiece barrel of compound instruments to replace the trapped air with cooler air from the outside world. However, having studied the webcam AVI videos of Damian Peach and Dave Tyler, obtained with 235- and 280-mm Celestron Schmidt-Cassegrains, it does appear that the sealed tube properties of SCTs do offer refractor-like closed tube advantages, with conventional Newtonian-type tube currents being virtually eliminated (as they are in a Newtonian at equilibrium). Maksutovs can be a different kettle of fish, though, and instruments over 20 cm aperture can have chronic cooldown problems as the curved corrector lens in a 25-cm instrument can weigh as much as the primary and the whole instrument can become a giant thermos flask. Maksutovs over 20 cm aperture without serious fan-cooling are best avoided. In 2005 Damian Peach acquired such an instrument (at great cost) and concluded it was totally unsuitable for planetary work. The optics were good, but the instrument stored heat like a bread oven. A Celestron 9.25 SCT, at an eighth of the price, proved to be a far superior planetary instrument as it cooled down much more rapidly!

Thermal problems are not just restricted to the telescope either. Thermally unfriendly observatories exist in abundance. Typically, these are brick-walled observatory domes with a massive concrete base, a narrow dome slit, and no ventilation system. Such structures can make observing the planets at high resolution virtually impossible after a sunny day. Planetary telescopes are far better if they are in the open air, so a run-off shed, run-off roof, or tarpaulin cover shelter are far preferable to a huge domed building. Finally, are you contributing to the thermal problems? Human beings give off a lot of body heat, so, when acquiring those webcam frames, stay well away from the telescope aperture.

Focusing

Although the rapid download speed of webcams makes planetary focusing much easier than with single-shot cameras, it is still a battle to focus a planetary image, especially when the planet is rippling and distorting at the whim of the atmosphere. In typical seeing conditions the planet looks like it is constantly being focused and de-focused. So how do you decide where the true focus point is? The best answer to this question will not sound very helpful; it is simply "try your hardest"! There are, admittedly, techniques used for focusing star fields that might be considered if you are desperate. The two most popular methods are the so-called Hartmann mask and the diffraction spike technique. In the former method, the telescope aperture is stopped down by two or more smaller apertures. For example, a 30-cm telescope might be covered by a 30-cm cardboard mask with two 7-cm diameter holes at the aperture edges. This will produce, in effect, two 7-cm telescope images. When the telescope is badly out of focus, two slightly overlapping images of the planet will appear. When perfect focus is achieved, the two images will merge. Unfortunately, while this method works fairly well for stars in low-resolution work, for planets we want the highest resolution we can get. The telescope resolution will be seriously hampered by the Hartmann mask, as, even when focused, the mask will cause horrendous diffraction effects. Worse still, planets at the sort of f-ratios used for imaging, especially faint planets like Saturn, will appear very ghostly when imaged through such a mask. The diffraction spike technique also only works well with bright pinpoint stars. Using this method, the secondary mirror support vanes of a Newtonian are used as an indicator of how sharp the focus is: a very bright star is imaged and the sharpness of the star's diffraction spikes are used to assess focus. (Artificial vanes can be placed over the aperture for non-Newtonians.) Unfortunately, this technique really does only work for long exposures with deep sky objects.

So, is there a solution to planetary focusing? Well, as a priority, a webcam planetary imager must have a motorized focuser. Even with the sturdiest telescope mountings the slightest touch of the astronomers hand on the focuser will shake the telescope. The best motorized focusers for amateur astronomy are made by JMI (Jim's Mobile Industries) and every top planetary imager I know uses one (Figure 3.12). Once a motorized focuser is in place, you can sit in a comfortable position at your PC screen, confident that the planet's oscillations are due to the atmosphere (and the telescope drive) and not your hand shaking the focuser. It goes without saying that the smoother the telescope drive, the better.

Figure 3.12. One of JMI's excellent motorized focusers. This one is an NGF DX designed for Newtonians. Focusing "shake" is eliminated with this design. Image: Martin Mobberley.

The globes of most planets appear as depressingly featureless fuzzy balls in the raw video stream from a webcam. Even Jupiter only shows two obvious features, namely the north and south equatorial belts. This can be quite demoralizing to the beginner, but do *not* worry, because raw images always look like this. However, with such ghostly features, what on Earth can you focus on? Mars and the Moon are really the only planetary bodies with sharp enough features to easily focus on. The Moon is a doddle, especially near the day/night terminator where contrast is high. With Jupiter, I use the Jovian moons as a focus reference. Yes, I know these are tiny discs and *not* point source objects, however, they are far better than anything else. On a typical night, with the webcam gain set high enough, Jupiter's moons will be easily recorded at 10 or 15 frames per second with a 25-cm reflector. However, the moons, even when focused, will rarely appear as small discs. They will appear as eggshapes, blobs, spiders, and even stubby lines as the atmosphere wreaks havoc on the incoming light. The best one can do is to spend a good 10 minutes, just moving in and out of the focus point and developing an educated "gut feeling" for where the best focus position is. While this may seem rather unscientific, well, to be honest, it is! However, planetary imaging is a bit of a black art. JMI moto-focusers have the option of an extra feature known as digital position readout. Typically, this tells you where the focuser is within 0.01 mm. This can be a useful aid when trying to remember when the planet looked sharpest. Getting back to the subject of Jupiter's moons: is there always one within a reasonable distance of Jupiter, I hear you cry. The answer is, invariably, yes. By creeping Jupiter east or west until it just leaves the webcam PC window you will almost always spot a Jovian Moon if the webcam gain is high enough. The closest Galilean Moon Io never strays more than 3.5 arc-minutes from Jupiter and with three other bright moons to consider you will never be short of focus targets. Of course, when seeing really is excellent you will be able to focus the Jovian moons into tiny discs. But this might only occur a few times per year. Fortunately, Saturn has one excellent feature to focus on: the rings. Specifically, the gap between the A and B rings, the Cassini division, is a focusing gift from heaven. Focusing is *not* something that can be rushed. After a while you develop a second sense of when something is focused "as well as it can be" and, at that point, you can start saving your webcam video to the PC hard disk.

Another crucial point here is the following one: does your motorized focuser actually allow you to increment the focus point in small enough bursts to exploit the very best seeing conditions. I have already mentioned that digital readout can record the focuser's position to within 0.01 mm, but it is often quite tricky to actually give a short enough "jab" on the focuser key pad to move it this small a distance. The increments needed for focusing a planetary image accurately are microscopic when you are working at Newtonian f-ratios. With an f/10 Schmidt-Cassegrain, a short jab on a motorized focuser fitted to the back of the telescope will just be precise enough. But with an f/5 Newtonian it will not, one jab will take you from inside the optimum focus position to well outside. A 25-cm f/5 Newtonian will, in theory, resolve about half an arc-second. This equates to 3 microns at the Newtonian focus. (A micron is 1/1000th of a millimeter.) This, in turn, equates to 15 microns of focusing tolerance at the same focus, or 0.015 mm. It makes no difference whether a Barlow lens follows the focusing point, unless you can arrange for the focuser to be *after* the Barlow lens. Even the best motorized focusers struggle to position the focuser to this accuracy and, even if they can, you literally need to jab the buttons for an imperceptible period. In my case, I contacted the focuser manufacturers and they told me which resistor I could change to make the focusing increments smaller. The only practical alternative would be to spend even more time jabbing back and forth across the focus point until you are happy, or switch to a telescope with a much longer nominal f-ratio. Schmidt-Cassegrains have a distinct advantage here if you add a motorized focuser to the drawtube where the light cone is a gentle f/10.

There is one other point I would like to mention in the context of focusing, and it adds an extra sledgehammer-type weapon to the planetary imager's arsenal. Essentially you need a lot of hard disk space! If you cannot determine exactly where the focus point is on the raw image, your chances of hitting it are increased if you take as many webcam videos as possible. Between each imaging run you should have another stab at focusing, because 1) you may get lucky; 2) the telescope tube may have contracted as the temperature dropped; and 3) the seeing might have improved, making focusing easier. When you can check the recorded runs indoors, at leisure, you may well conclude that one of them is much better than the rest. When images from this best run are stacked and processed, a superb image may result. However, bear in mind that you do need a large hard disk. Planetary imaging runs can be 1 or 2 Gigabytes in size. If you take a dozen during the night that is a lot of hard disk space.

Above and beyond all these high-resolution considerations are the issues of image processing wizardry. I will have much more to say about these in Chapters 8 and 9.

Before we leave the issue of high-resolution considerations it is time we indulged in the never-ending debate as to what constitutes the best high-resolution instrument.

Ultimate Planetary Telescopes

The debate is as old as the hills. What is the ultimate planetary instrument? Is there such a thing at all and what, exactly, do we mean by "ultimate"? I have touched on this subject already when discussing the relative merits of my own Newtonians and the long-focus Newtonian. One way of answering this question is simply to look at what telescopes the world's leading planetary imagers use. However, when we do this we just see a reflection of the market share of different telescopes. Damian Peach, arguably the world's finest planetary imager, and certainly the keenest, has achieved great success with Celestron's 9.25 (235 mm) and 11 inch (280 mm) Schmidt-Cassegrain telescopes, operated at f. ratios up to f/40 (Figures 3.13 and 3.14). In his view, the optics of Celestron's SCTs are superb, the only

Figure 3.13. Arguably the best value-for-money planetary telescope, the Celestron 9.25's are renowned for their planetary performance. A slightly longer f-ratio primary (f/2.5 not f/2.0) and lower "magnifying" secondary (4x not 5x) is thought to be responsible. Image: Damian Peach.

Figure 3.14. A well-thought-out and well-used planetary telescope. Dave Tyler's 280-mm Celestron 11 and 150-mm Intes Maksutov-Cassegrain mounted on his home-made German Equatorial mounting. The smaller telescope is invaluable for aligning the main telescope on the tiny planetary field. Image: Dave Tyler.

disadvantage with the design being that, like all SCTs, collimation does not stay fixed. A nightly check and tweak is always required. Damian has used both Celestron and Losmandy drives to support his Celestron Optical Tube Assemblies. The SCT is the most popular serious amateur telescope design there is, but it is not obvious why this should be. It is, optically, a halfway house in telescope design, not especially suited to either high-power or wide-field observing. However, it does have two huge advantages: SCTs are extremely compact and mass production has made them affordable. A compact telescope is user-friendly, portable, and light-weight, so only a modest drive is needed to mount it. Also, with tube assemblies weighing 15 kg or less (except for the largest models) the instrument can be stored indoors, so no observatory is required. Despite Damian's status in planetary imaging he has *never* owned an observatory. His telescope is set up outside every single night, even if there are only tiny cloud gaps. This nightly hassle would be impossible with a less compact instrument.

Optical Quality

A word or two here about optical quality may be appropriate. Manufacturers quite often use the term "diffraction limited." This means that the optics are good enough that they are only limited by the laws of physics, and the property known as "diffraction" limits the resolution of all instruments, whether optical or radio telescopes.

The larger the mirror is in relation to the wavelength, the better the resolution you will get. To be "diffraction limited," an optical telescope needs to have optics that, as a minimum, deviate from the perfect shape by no more than a quarter of the wavelength of light. In other words, from the deepest valley below the perfect curve, to the highest peak above it, should be less than a quarter of a wavelength. Green light has a wavelength of about 550 nanometers, so a quarter-wave is about 140 nanometers. This is what the term quarter-wave PV means (PV = peak to valley). If the mirror is less accurate than this, then planetary detail will get noticeably soft and mushy. A quarter-wave PV equates to plus or minus one-eighth-wave surface accuracy. Another factor here is RMS, or root mean square. This is an indicator of how smooth the telescope mirror is, on average, rather than just taking account of the most extreme peaks and troughs. Typical commercial Schmidt-Cassegrain mirrors tend to be one-quarter-wave PV or one-sixth-wave in really good examples (such as Celestron's C9.25 models). In other words, they are diffraction limited. Commercial lemons occasionally crop up though, with half-wave PV optics. Commercial mass-produced optics usually have RMS figures of 1/20th to 1/30th wave, but a really good set of Newtonian optics may have an RMS of 1/40th wave and a PV figure of 1/8th or better. These are optics to be proud of.

In the 1970s and 80s, SCTs had quite a poor reputation for optical performance. One factor often cited by SCT critics was the size of the secondary mirror, i.e., the central obstruction. The traditional advice of experts was that a telescope should have an obstruction less than 20% the diameter of the main mirror, if diffraction effects were not going to degrade the view. The effect of any obstruction in the telescope light path is to reduce the contrast at the telescope limit. In an unobstructed telescope the vast majority of the light from a point source like a star ends up in a central point with the remainder distributed in a series of rings, becoming increasingly fainter as you move out from the central point. As you increase the central obstruction, the intensity of the central point is reduced while the intensity of the rings increases. How does this affect the planetary performance? Well, without going into complex analyses involving mathematical terms like "MTF" (modulation transfer function) and Strehl, visual observers often quote a rule of thumb that a telescope of aperture x with an obstruction of aperture y reveals subtle planetary features as if it were an instrument of aperture $(x - y)$. In other words, a 30-cm SCT with a 10-cm obstruction will behave like a 20-cm apochromat refractor (a refractor with no visible color aberrations). Most planetary observers think this rule is about right, although when observing the Moon, where there is so much contrast available, a large central obstruction is much less of a problem. Of course, there is another factor here. A 30-cm instrument with a 10-cm obstruction still has twice the light grasp of a 20-cm refractor. With more light there is a higher signal-to-noise ratio in the webcam image (for the same image scale) and you can choose to have a shorter exposure to better freeze the seeing turbulence. But there are downsides to larger apertures in that you have a heavier less-friendly instrument, one with more thermal mass and one that will rarely be able to resolve to its theoretical limit anyway. If you want to really simulate a telescope's optical performance, I would strongly advise downloading another excellent piece of freeware written by Cor Berrevoets. This software package is called Aberrator (available at http://aberrator.astronomy.net/).

When I first came into amateur astronomy I was convinced that bigger was better. I wanted the largest telescope I could acquire. Eventually, I owned a massive 49-cm aperture Newtonian. However, as the years have gone by it has become more and more apparent to me that a user-friendly telescope is the best telescope to have. Specifically, a quality user-friendly telescope, with a reliable drive and good optics, that can be easily collimated, is what is required. This is especially true in planetary observing where the atmosphere rarely allows features much smaller than 0.5 arc-seconds to be seen. Over the years I have consulted many leading planetary observers and imagers on what they think is the largest useful aperture that you need. Many of these observers have used extremely large telescopes at high altitudes with mirrors up to 1 meter in aperture. The general consensus is that a 25-cm telescope will show you all that you need to see on 95% of nights, even on a planet that is at 50 or 60 degrees altitude. Yes, there are freaky nights, maybe once a year, when a 30-, 35-, or even 40-cm instrument could benefit the observer, but the user-unfriendliness of such instruments will probably have a negative effect on the observer's enthusiasm. In addition, the thermal properties of such large instruments and the likely quality of their optics will work against such telescopes giving any benefit. One of the "holy grails" in planetary observing is seeing the elusive Encke division in Saturn's A ring. The elusiveness of this feature, even in large apertures, is a testimony to the idea that large apertures rarely have any benefit to the planetary observer. The Encke division (not to be confused with the Encke minimum, which is simply a subtle shading effect between inner and outer parts of the A ring) is a feature that resembles a human hair, right on the limit of visibility in amateur instruments. It is 1/20th of an arc-second across, so well below the resolution of instruments less than 2 meters in aperture. However, it can be glimpsed, even in instruments as small as 15 or 20 cm, because it causes a contrast drop in that part of the A ring. Despite this fact, it took a night of perfect seeing on the 36-inch Lick refractor in January 1888 to confidently record the feature for what it was. That definitive and undisputed observation was made by James Keeler. The Encke division is still an elusive feature, even in the webcam era. However, in perfect seeing it has easily been recorded, at Saturn's ansae (the east and west ring tips) and with the rings wide open, with 20-cm instruments and a webcam.

I think I have provided quite a bit of evidence here that large 35- and 40-cm aperture instruments are just not needed for planetary observing. Indeed, their user unfriendliness can be off-putting. But is there actually a perfect planetary telescope? Well, all I can do is list a few examples of commercial telescopes that have been used to good effect by some of the world's keenest webcam and CCD imagers.

As I have already mentioned, Damian Peach has achieved staggeringly good images with Celestron's 11 and 9.25 inch SCTs (28 and 23.5 cm in metric). The C 9.25 has something of a legendary status as the best planetary SCT you can buy. Part of the reason for this is that it has a longer f-ratio primary mirror than other SCTs (f/2.5 compared to f/2.0), which significantly improves the chances of there being less optical defects and aberrations. It is also of an ideal aperture for exploiting the best the atmosphere can offer, while packaged in a lightweight, low thermal mass optical tube. I know a lot of observers who have owned C 9.25s and none have been disappointed. The French planetary imager Thierry Legault has achieved very impressive results with a 30-cm Meade LX200 mounted on a Takahashi German Equatorial mounting, thus combining good optics with a superb, smooth drive. SCTs only have two main disadvantages: the "flopping" of the primary mirror out of collimation and dew formation on the corrector plate (when it occurs on the inside of the corrector, in very damp conditions, this can be especially frustrating as it leaves a haze long after the dew has gone). One of the best-known "connoisseur" instruments for planetary imaging is the Takahashi Mewlon 250 Dall-Kirkham Cassegrain. Unlike with SCTs, Dall-Kirkham telescopes have no corrector plates, so their optics are exposed to the night air and cool down more quickly. The Mewlon 250 also has a detachable rear mirror cell plate, enabling the primary mirror to cool down very swiftly to the night air. Takahashi's Mewlon 250 is a superb performer, although a lot more expensive than an SCT of the same aperture. The Singapore observer Tan Wei Leong has produced exquisite results with his Mewlon. Veteran Florida amateur Don Parker, best known for his massive 40-cm f/6 Newtonian, has a Mewlon 250 too. In the U.K., Orion Optics produces an excellent planetary Maksutov called the OMC 200, which is well worth considering (Figure 3.15). Their newest open-tube Maksutov Cassegrains (the OMC 300/350 models) are intended both for planetary and deep sky work. For those with a love of Russian telescopes, the company Intes-Micro produces some superb modest aperture Maksutov-Cassegrain and Maksutov-Newtonian instruments (Figure 3.16). The Maksutov-Newtonian is a particularly interesting design, combining the quality of a slow Newtonian primary mirror with a corrector plate to reduce Newtonian aberrations. Such an instrument has a sweet spot of a similar size to a long-focus Newtonian. However, just as important as the optical specification is how an instrument performs in nightly use. An instrument may have perfect optics, but, if, for example, the collimation process is almost impossible, it is as useful as a chocolate teapot, an inflatable dartboard, or a rubber pick-axe. Some of the larger Maksutov-Newtonians in circulation have excellent Russian Intes-Micro optics mounted in optical tubes made by less competent dealers . . . buyer beware! The website www.cloudynights.com is a very useful site to visit when assessing what telescopes are actually like to use, in reality, away from all the advertising hype.

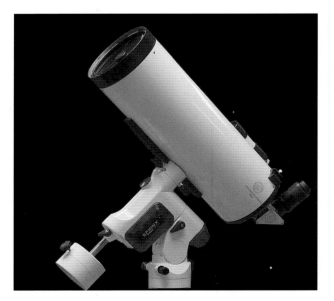

Figure 3.15. Orion Optics excellent 20-cm f/20 Maksutov, the OMC 200. A superb and compact planetary telescope. Image: Orion Optics.

Figure 3.16. Another excellent planetary telescope: Intes Micro's MN78, an 18-cm f/8 Maksutov Newtonian with a tiny secondary obstruction and closed tube. Image: Jamie Cooper.

If you have very deep pockets, a 20- or 25-cm apochromatic refractor is, perhaps, the ultimate planetary observer's status symbol. But only consider this option if you are happy spending $20,000 or $30,000 on the optical tube alone. A number of specialist large refractor companies will be only too happy to empty your wallet and provide you with such an instrument.

From my own perspective, as previously mentioned, my favorite planetary instrument is currently my 250mm f/6.3 Orion Optics Newtonian (Figure 3.17). It has superb, easy to collimate optics (1/45th wave RMS), a very convenient eyepiece height, and a cooling fan at the mirror back. The 1-mm-thick tube cools to ambient temperature quickly and yet is rigid enough to maintain the system in collimation. The telescope also features an unusual double circle secondary holder that eliminates diffraction spikes around bright stars (Figure 3.18). The only modification I have made to it is to add a motorized JMI focuser. The optical tube only weighs 11 kg. However, even this system is not quite perfect. The Sphinx mount that the telescope was supplied with is right on its weight limit and long-tube Newtonians are quite susceptible to wind-shake and vibration, especially if the observer is changing filters while the system is in use. A Newtonian of a certain weight needs a much heftier mounting than a Schmidt-Cassegrain of the same weight to avoid rigidity problems. One solution to this sort of rigidity problem, and far cheaper than a bigger mounting, is the so-called "Hargreaves Strut." This modification, named after the historical British amateur astronomer, telescope-maker, and wartime BAA president F. James Hargreaves, employs a strut (with universal joints) between the end of the telescope tube and the German equatorial counterweight arm, to brace the system.

Figure 3.17. The author and his 250-mm f/6.3 Orion Optics SPX Newtonian. Image: Martin Mobberley.

Figure 3.18. The unusual secondary mirror (spider) on the author's Orion Optics Newtonian. This design eliminates diffraction spikes around bright stars. Image: Martin Mobberley.

Undoubtedly the best "ultimate planetary telescopes" of all are those designed and built by the amateur astronomer for his or her own use. ATM, or amateur telescope making, has declined as a hobby throughout the 1980s and 1990s and into the 21st century, largely due to the mass-production of amateur telescopes and the pressures of modern working lives. There is another factor that is relevant too. Many ATM fanatics are far more interested in building telescopes than actually using them. ATM is their all-consuming passion, not observing. With me it is the opposite. The telescope is the tool to get the job done. My hobby is imaging astronomical objects, *not* building telescopes and I would rather spend time at the eyepiece or working on images than working on a lathe. However, while the average amateur with a job and family simply does not have the time to make telescope optics or a telescope drive from scratch, the telescope tube is something that lends itself well to customizing. The Newtonian is the easiest tube to modify and dramatic enhancements can be made simply by adding cooling fans, making a thin vane secondary support system, and, most important of all, making the system easy to collimate and rigid. As I have previously mentioned, in days gone by, amateur astronomers used to make Newtonian telescope tubes from mahogany rather than aluminium to avoid the metal tube radiating more tube currents into the light path as it cooled down. Often, the inside of the tube was lined with cork to further reduce tube currents. If the Newtonian tube is only fractionally wider than the mirror, tube currents when cooling are more likely. A Newtonian tube should have a radius a good few centimeters wider than the mirror radius. However, we have seen that as a tube becomes more "open," dew is more likely to form on the primary and secondary mirrors. This can easily be removed by a hair-dryer, but preventing it from forming in the first place is better. A dew heater band will cause permanent heat turbulence in the light path whereas the momentary blast from a hair-dryer will dissipate away. Of course, as soon as the primary or secondary have cooled, dew begins to form again! While discussing custom-built planetary Newtonians I have to mention the excellent long-focus planetary Newtonian of York amateur Mike Brown. This is shown in Figure 3.19, along with Mike's exquisite mirror cell in Figure 3.20. Such a telescope will stay collimated for life because the diffraction-limited sweet spot is so large.

Because the planetary webcam user's telescope does not need to be large, a simple user-friendly observatory is well within the abilities of even the most hopeless DIY enthusiast to construct. For my 250-mm Newtonian I thought long and hard about what the best, most user-friendly, and least obtrusive telescope shelter might be. I came up with the design shown earlier in Figure 3.17.

Because precise polar alignment is not essential for planetary observing, the telescope glides out on rails from a kennel-like structure attached to the southeast-facing wall of the house. Everything can be up and running in a matter of minutes and the telescope is the heaviest structure that needs to be moved.

Before I finish this chapter, I would just like to say a few words about bad optics. Figure 3.21 shows a Zygo interferometer plot of a 40-cm mirror that was advertised as being accurate to 1/10th wave PV but was actually only accurate to half a wave! This mirror was made by a backstreet mirror-maker who claimed his mirrors were far superior to mass-produced ones. In fact, they were far inferior. Fortunately, such con men do not survive for long as they are soon found out with the power of a Zygo interferometer.

Figure 3.19. The 250-mm f/9.3 Newtonian of Mike Brown from York, U.K. The instrument has a diffraction limited field of almost 18 mm diameter at the Newtonian focus. Image: Mike Brown.

Figure 3.20. A well-designed mirror cell, made by Mike Brown from York, U.K., for his 250-mm Newtonian. Note the three-triangle/nine-point suspension system and the open frame allowing air circulation. Considerable attention has been applied to making the design concentric with the tube mounting holes. Image: Mike Brown.

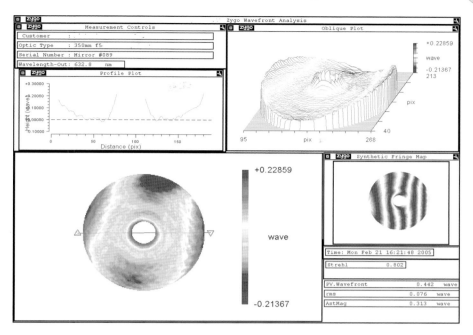

Figure 3.21. A competitors mirror advertised as 1/10th wave PV actually turned out to be 0.442 wave, 4.4 times worse than advertised. Star images in the instrument were noticeably soft and planetary views very disappointing for a 350-mm instrument. Image: Orion Optics.

Finally, before we leave the chapter on high resolution, I would like you to have a look at Figure 3.22. This is an image, taken by Damian Peach, of the intra- and extrafocal diffraction patterns seen in a badly designed 25 cm f/12.5 Maksutov. The optical tube, despite costing $12,000, was so badly ventilated that it could never deliver quality results. Indeed, side-by-side with a 23.5-cm Celestron SCT, costing one-eighth (!!) the price, its performance was pitiful. Price does *not* necessarily mean performance!

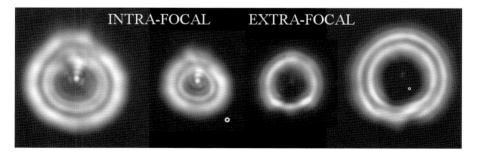

Figure 3.22. Thermal problems in a $12,000 Maksutov-Cassegrain are obvious when one examines the intra- and extrafocal diffraction patterns. The distorted, fluted pattern is characteristic of heat seeking a way out. Image: Damian Peach.

CHAPTER FOUR

Planetary Imagers Worldwide

Despite the ease with which excellent planetary images can now be obtained, planetary imagers are rather thin on the ground among the population. Across the world there are only a few dozen regular planetary imagers whose results are good enough to be used by professional astronomers. Undoubtedly the main reason for this is simply the rarity of good, stable, atmospheric seeing coupled with clear skies, along with the dedication and knowledge required to get world-class results. However, a few amateurs do persevere and it can be fascinating to study their methods. The world's most experienced planetary imager is Donald Parker of Coral Gables, Florida. Don goes right back to the days of planetary photography, and his photographs of Mars, Jupiter, and Saturn were, for many years, in a league of their own. To get some idea of what results were attainable by amateurs using film in the 1980s, check out the Parker-Dobbins-Capen book *Observing and Photographing the Solar System* published by Willmann-Bell in 1988.

Parker's original photographic telescope was a 32-cm f/6.5 Newtonian, but his main instrument for many years has been a massive 40-cm f/6 Newtonian, shown in Figure 4.1. However, he has also used a 40-cm f/10 Meade LX200 (later mounted on a Paramount mounting) and a 25-cm Takahashi Mewlon is used on nights where the 40-cm's aperture and potential resolution is not required. Don Parker's Coral Gables location near Miami, Florida, must be a big advantage. Florida is a humid area of the southeastern U.S., not far from the Caribbean; the area is renowned for good stable seeing, especially at sites near the sea, where a laminar air flow can prevail. One renowned telescope and eyepiece manufacturer once stated that seeing in the Florida Keys was so good that he did not need a laboratory with an artificial star to test his equipment. Also, at 26° north, the planets can attain altitudes that some of us (me included) can only dream about. At their highest declinations, Mars, Jupiter, and Saturn pass virtually overhead. Of course, there is the downside of a regular battering from hurricanes! Although Parker's Florida

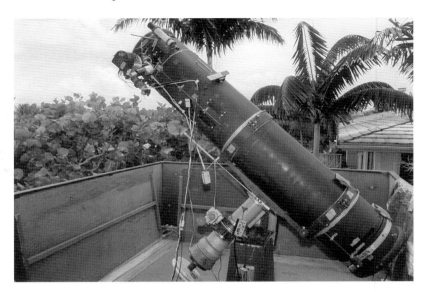

Figure 4.1. The 40-cm f/6 Newtonian of Donald Parker of Coral Gables, Florida. Possibly the most productive planetary telescope of all time and, rarely, it spans the eras of photographic, CCD, and webcam imaging! Image: Don Parker.

location and latitude (as well as his colossal dedication) placed him in a different league from other amateurs in the 1980s, the CCD and webcam era has eroded that advantage somewhat. Poor seeing (prevalent at most locations and when planets are low down) was an inpenetrable barrier in the era of planetary photography, but webcams, and their ability to freeze the seeing, have significantly reduced that advantage. The effects of atmospheric dispersion can be digitally reduced by realigning the color layers too. Nowadays, Parker uses his large-aperture Newtonian to give him better signal-to-noise images than the competition. Using just a single Barlow lens at f/6, to give an f/14 focal ratio with his 40-cm instrument, Parker's system gives a fairly modest focal length of 5,600 mm when used with a ToUcam Pro. In the last year he has started using an ATiK color webcam with 7.4 micron pixels at f/22, giving an image scale of 0.17 arc-seconds per pixel.

In the late 1980s, Japanese amateur Isao Miyazaki produced planetary photographs to equal Don Parker's best, using a truly massive and professional quality 40-cm f/6 Newtonian housed on the roof of his apartment block on the island of Okinawa. Miyazaki arrived on the scene at the same time that ultrafine-grain Kodak 2415 film helped improve image quality. Okinawa is at latitude 27° north. Miyazaki's Newtonian is now housed in a dome on top of his Okinawa house, as shown in Figure 4.2.

In the late 1990s two amateur CCD imagers raised the bar to an unprecedented level. Frenchman Thierry Legault, using a 30-cm Meade LX200 and a Hi-Sis 22 CCD camera, achieved astonishing results and was, perhaps, the first of the modern imagers to truly appreciate all of the factors involved. Perfection in the fields of telescope collimation, focusing, stacking dozens of images, and image processing were

Figure 4.2. The magnificent 40-cm f/6 Newtonian of Isao Miyazaki of Okinawa, Japan. Miyazaki is one of the world's most experienced planetary observers. The telescope was built by Yasuyuki Nagata in 1988. The optics are by Ichirou Tasaka. Image: Isao Miyazaki.

his hallmark. His images, revealing Saturn's Encke division, astounded both amateur and professional observers.

As the 20th century came to a close another world-class imager emerged from a most unlikely location: Damian Peach from the U.K. (Figure 4.3). Although U.K. seeing is probably no worse than for most other locations in the world, the southern U.K. is at 50–52 degrees north and the vast majority of nights are cloudy. However, at the turn of the 21st century, both Jupiter and Saturn were at a very high northerly declination and Damian took full advantage of the new CCD technology and techniques from his seventh storey apartment location in Norfolk. After moving briefly to Kent, Damian decided that he just had to leave the U.K.'s cloudy skies behind and, in 2002, he briefly moved to Tenerife so that Jupiter and Saturn would be almost overhead. Damian switched from using a 30-cm LX200 to a 28-cm Celestron 11 at this time and the results Damian obtained in 2002 and 2003 raised the standards even higher than those set by his predecessors.

Figure 4.3. Damian Peach, arguably the world's keenest top-quality planetary imager. In the last few years, Damian raised the quality bar and set new standards of planetary imaging. Damian is shown here next to Patrick Moore's famous 38-cm f/6 Newtonian at Selsey, UK. Image: Martin Mobberley.

Moreover, his wealth of experience, from observing virtually every clear night, in the U.K. and in Tenerife, revealed some interesting information. For a start, Damian concluded that Tenerife was far from ideal in respect of atmospheric seeing. Indeed, he experienced a few nights when the planets were unbelievably blurred, mainly when he was on the leeward side of Mount Teide and turbulent air rolled down his side of the mountain. Tenerife was definitely a superior site, but largely from the point of view of the number of clear nights and the altitude of the planets and *not* from vastly superior seeing. The other huge advantage of course was in the comfort of the night time observing temperatures, a far cry from the sub-zero icy winter misery Damian was used to suffering from in the U.K. Damian has achieved extraordinary success with Celestron 11 and Celestron 9.25 Schmidt-Cassegrains and in many ways is the successor to the U.K.'s Terry Platt, who pioneered U.K. CCD imaging with his company, Starlight Xpress. Before Terry, the legendary mirror-maker and planet-imager Horace Dall was the U.K.'s top planetary photographer. Amazingly, two of the world's finest planetary imagers now live within a mile of each other near Loudwater in the U.K., as Dave Tyler (Figure 4.4), now the proud owner of Damian's original Celestron 11, is also producing fantastic planetary images.

Two amateur astronomers based in the Far East, who specialized in planetary imaging, also emerged around this time. Eric Ng of Hong Kong and Tan Wei Leong of Singapore both observed from high-rise apartment buildings using relatively modest equipment. Tan Wei Leong was obtaining excellent images with a

Figure 4.4. U.K. imager Dave Tyler with his Celestron 11 (formerly owned by Damian Peach) and 15-cm Maksutov. Image: Dave Tyler.

Celestron 11 before he switched to an equally superb Takahashi Mewlon 250. He also used a 40-cm Cassegrain reflector at Singapore Observatory during the Mars opposition of 2003.

Eric Ng (Figure 4.5) uses simple 25- and 32-cm f/6 Newtonian reflectors to achieve his stunning results. His reflectors employ mirrors by the U.S. optical worker William Royce, an astronomical mirror-maker of high renown.

Ed Grafton, of Houston, Texas, (Figure 4.6) has been one of the world's leading planetary imagers for many years, although he has largely resisted the webcam revolution, preferring to concentrate on using his SBIG ST5c CCD camera. In excellent seeing the advantage of a webcam is largely negated as its main strength is in its ability to take thousands of frames, such that a fraction of the frames will be good enough to stack up to produce a less noisy composite. If your telescope was in space, like the Hubble space telescope, a single image would be enough and it would be a true snapshot of the planet in time (rather than a composite over several minutes). Astronomical CCD cameras are, generally, more quantum efficient than webcams and their capability for longer exposures, plus their cooler temperatures means that individual frames are much less noisy. Thus, a good image from a cooled CCD camera may consist of a few dozen or a hundred frames, rather than hundreds and thousands of frames. This has been Ed Grafton's approach with his Celestron 14 and the technique has worked well. Like Don Parker, 1,000 miles to the east, Ed is situated in the southern U.S. At latitude 30° north, the planets usually transit at a very respectable altitude.

Christophe Pellier (Figure 4.7) is another amateur worthy of a mention. Observing from France, Christophe has obtained exquisite planetary images with surprisingly

Figure 4.5. Eric Ng of Hong Kong with his 32-cm f/6 Newtonian, featuring Optics by William Royce and an Astrophysics mounting. Image: Eric Ng.

Figure 4.6. Ed Grafton (left) and Don Parker (right) pose in front of Grafton's 35-cm Celestron 14 at Grafton's Houston observatory. Image: Ed Grafton.

Figure 4.7. The French observer Christophe Pellier, one of the world's most active planetary imagers, with his modest, but productive, 180-mm Newtonian. Image: Christophe Pellier.

small apertures, including a 180-mm aperture Newtonian. His fellow countryman, Bruno Daversin, has adopted the opposite approach, using a massive 60-cm Cassegrain at the Ludiver facility (L'Observatoire Planetarium du Cap de la Hague) to obtain the sharpest lunar images ever obtained from Earth. In passing, it is worth mentioning that France seems to have had a substantial number of top planetary observers over the years, even if not all of them were actually French by birth. In the photographic era, Professor Jean Dragesco dominated the scene and produced an excellent book in 1995 entitled *High Resolution Astrophotography*. Dragesco observed from a number of excellent sites during his astrophotographic career, but has now retired in southern France. Another French astrophotography legend of the 1970s and 80s was Christian Arsidi, who obtained excellent lunar results using a 250-mm Takahashi Mewlon Dall-Kirkham Cassegrain and a 310-mm Cassegrain. In the 1980s and 90s, another Parisborn amateur, Gerard Thérin astounded amateur astronomers with his remarkable lunar pictures taken with a 203-mm Celestron Schmidt-Cassegrain. For a time, Arsidi and Thérin collaborated. Finally, Georges Viscardy (who this author visited in 1989) built an impressive observatory in the hills north of Nice featuring a massive 51-cm Cassegrain and a 30-cm f/7 Newtonian. Viscardy published a truly massive photographic lunar atlas during the 1980s, which was of the highest photographic quality.

Just before the webcam era, another Florida amateur, Maurizio Di Sciullo, using a superb 25-cm f/8 Newtonian, became renowned for routinely imaging details on the Jovian satellites! I remember the first time I saw one of his Jupiter images, showing markings on the satellites Ganymede. I was astounded. Maurizio used a monochrome Starlight Xpress CCD camera plus a filter wheel for his images. Figure 4.8 shows Maurizio's excellent 250mm f/8 Newtonian. Maurizio raised the bar to unprecedented levels as the 20th century came to a close.

In the webcam era more planetary imagers have emerged, but the highest quality imagers are as rare as ever. In Portugal, Antonio Cidadao specializes in filtered planetary imaging, such as methane band images of Jupiter. Such work benefits from large apertures and Antonio uses a 35-cm LX200 Schmidt-Cassegrain. In Italy, Paolo Lazzarotti has obtained many exquisite lunar and planetary images and in Spain, Jesus Sanchez has been that country's top planetary imager for many years. In the Phillipines, Chris Go has achieved stunning results with a 20-cm Celestron, whereas Jim Phillips of the U.S. uses 20- and 25-cm TMB apochromats. In Japan, Toshihiko-Ikemura is, arguably, the world's most prolific Mars imager.

Many of these observers live in favorable locations, but not all. However, if one lives in a cloudy, high-latitude location all hope is not lost. A trip abroad with relatively modest equipment can yield excellent results, as we will see in the next chapter.

Figure 4.8. A telescope that raised the planetary imaging bar. Maurizio Di Sciullo's 250-mm f/8 Newtonian featured excellent optics, a cork-lined tube, and a highly reflective outer skin. Image: Maurizio Di Sciullo.

CHAPTER FIVE

Have Webcam, Will Travel

Planetary observers at high northern and southern latitudes are at a huge disadvantage when compared to their counterparts nearer to the equator. Take the situation of this author, for example. I live at 52° north and so the highest the planets ever get above my southern horizon (when they are at + 23° Declination) is 90° − 52° + 23° = 61°. Most of the time the planets are much lower, of course, typically, an altitude of 40° is better than average for planets from the U.K. But when the planets sink deep into negative declination territory, all hope is lost. Mars, at its perihelic oppositions, is especially bad from the U.K. as it rolls along the south horizon barely clearing 20° on the meridian. While narrow-band filters, especially in the deep red, can be used to salvage a half-decent image this is no substitute for the planet being high up. Low altitudes not only cause dispersion, spreading every bright feature into a rainbow of color, they also mean poorer seeing (as far more turbulent air is being encountered) and serious degradation in the blue end of the spectrum.

However, there is a solution to this problem and one especially compatible with using a webcam, a laptop, and a modest telescope. That solution is traveling abroad for a couple of weeks (or longer) to coincide with the planet's opposition period. Air flights have never been cheaper and planetary equipment has never been as portable. But where are the best places to go and what should you take?

Holiday Destinations

Of course, a lot depends on where you actually live; an amateur astronomer in the U.S. is unlikely to prefer a Mediterranean destination to a Caribbean one unless he or she wants to combine a European/north African sightseeing trip with a bit of astronomy. The requirement is, of course, to head closer to the equator so that the planet is higher up. Also, a site that enjoys clear skies much of the time is essential, as is a site renowned for its good seeing. As we are planning on heading nearer to the equator we should have little concern about nighttime temperatures, unless we are fortunate enough to have access to a high altitude site. On this point it is surprising how cold it gets once you have ascended a few thousand feet above sea level. If you are planning a high-altitude session, also bear in mind that most people will become quite breathless if they are working hard at above 7,000 or 8,000 feet. Sadly, there have been a few fatalities over the years when unprepared explorers and amateur astronomers have ascended to high altitude at night, become cold, confused, and disoriented, and then attempted to drive back down icy mountain roads with an icing up windscreen. Never overestimate your ability to perform in freezing temperatures and at altitude, and make sure wind-proof thermal clothing is packed and suitable (Figure 5.1). For planetary observing, where transparency is not critical, the pleasant nighttime temperatures at sea level, near the tropics will result in far more being achieved before fatigue sets in (Figure 5.2).

Figure 5.1. Dave Tyler models lightweight, but warm, clothing for those long, cold nights outdoors. Note the hairdryer for removing dew: an essential night-time accessory in most climates. Image: Dave Tyler.

Figure 5.2. Damian Peach, when temporarily resident in Tenerife, enjoyed much warmer evenings than in the U.K. Note the modest and transportable equipment.
Image: Damian Peach.

Locations that are close to the sea are frequently favored as far as good seeing is concerned, providing one is not on the leeward side of an island under the turbulence cascading down the central hills or mountain. It is far better to be on the side of the island where a laminar air flow coming in off the sea hits the land i.e., on the windward side. However, in truth, it is far better to be in a situation where there are zero winds from ground level to high altitude. This stable situation virtually guarantees good seeing. The sea is a huge reservoir of heat and so tends to prevent huge day- and nighttime temperature variations, ideal for good seeing. In a perfect world, the dream planetary imager's island would have the same sea, air, and land temperature, day or night. If you are surrounded by water this helps considerably in that regard.

Florida, especially along the coastal region, is renowned for good seeing. The southern tip almost reaches down to a latitude of 25° north; a vast improvement if you come from the northernmost U.S. Go further south and you come to the islands of the Caribbean; beautiful tropical island paradises and with the planets even higher up and renowned seeing. What could be better?

Of course if you are keen to search out the very best sites on Earth, for good seeing and clear skies, you need do nothing more than search out where the best professional observatories are. Professional astronomers spend years searching out the best sites, for seeing, low rainfall, low humidity, and darkness, and most of these high-altitude sites are excellent for planetary work, except from the viewpoint of their accessibility, low oxygen, and cold night temperatures. (Incidentally, high humidity is often good for planetary conditions, but low humidity is indicative of a dry, rain-free and cloud-free site.) ESO's Cerro Paranal Observatory in the Chilean Atacama desert typically experiences 0.6 arc-second seeing but, in Europe, the famous Pic du Midi Observatory in the French Pyrenees also fares well. Jean Dragesco, in his excellent 1995 book *High Resolution Astrophotography*, pointed out that the climates of a number of countries in equatorial Africa were excellent for planetary photography, although a lot depended on precisely where the observer was located and what the season was.

For the northern European astronomer, the island of Tenerife is a cheap holiday destination with regular flights and a professional observatory established on the slopes of Mt. Teide. U.K. planet expert Damian Peach spent a year on Tenerife and secured some of the best Earth-based Jupiter and Saturn images during that time. However, as I have already mentioned, Damian did find that the seeing was far from excellent on many nights. The real advantage was the number of clear nights and the fact that the planets were so high up that atmospheric dispersion was not an issue (both Jupiter and Saturn were at a declination of +20° during that year, so they passed almost overhead from Tenerife). The neighboring island of La Palma is an even better site, although has fewer flights to it because it is not a major tourist resort in the category of Tenerife. The U.K.'s William Herschel Telescope is located at altitude on La Palma and the seeing from that site is exceptionally good, which is why it was selected. If you can travel to a site where a professional observatory is situated you can virtually guarantee that seeing conditions will be good, however, you will have to get permission first. Not surprisingly, the world's major observatories have strict regulations about the use of flashlights close to their facilities when they are trying to image stars of magnitude 27 or 28! If you live in the U.K. and regularly check the weather for signs of a high-pressure system or some good atmospheric stability you will not fail to spot that the Azores enjoy more than their fair share of high-pressure systems. Indeed, weather forecasters frequently refer to an "Azores high." In winter they seem to stubbornly refuse to settle over the U.K., and when they do they are frequently cloudy highs or highs west of Ireland feeding bitterly cold northerly air across the U.K. They seem to like to settle over the Azores themselves! Further down, and not too far from Tenerife, the island of Madeira has long been a favorite destination for planetary observers, even in the 1800s. Madeira has one of the most equable climates of any place in the Atlantic Ocean, with day (and night) temperatures at sea level, throughout the year, rarely dropping much below 20°C or rising much above 30°C. The British Victorian Astronomer Nathaniel Green made a special trip to Madeira in 1877 to observe Mars at its closest. Observing from the hills east of the capital Funchal (at an altitude of 1,200 feet) Green observed Mars on 26 out of 47 nights and described 16 nights as being either good, excellent, or superb. If Green could transport a substantial 33-cm reflector to Madeira in 1877, 21st century amateurs should, perhaps, not complain about carting a lightweight Schmidt-Cassegrain a similar distance.

Unlike the situation in 1877, the modern amateur only has to spend a few nights on the Internet downloading weather statistics to form a pretty good opinion of how suitable a site is for observing from. Indeed, these days it is very likely that other experienced amateurs can advise on the weather and seeing prospects for most obvious locations.

Equipment For The Trip

On a planetary imaging trip you will need to cart several thousand dollars worth of fragile equipment to your destination so, the first point to note is that you will need adequate travel insurance. Your basic insurance will almost certainly be inadequate. Also, when observing far from home, you cannot suddenly dash indoors to get some vital piece of equipment if a problem arises. Therefore, some forward thinking is highly advisable. Note well the bits and pieces you use at home before you fly abroad. Note very well the dodgy bits of your telescope that might "drop off" in transit and carry some spare parts if possible. Forgetting one vital item could mean you fail to secure any results on the best night. What sort of things am I talking about here? Well, for a start, the most vital observing accessory I have is a hairdryer; an absolutely essential piece of equipment in the nighttime U.K. climate. Even with the commercial dew heater equipment available, there are still nights when only a hairdryer will stop something misting over. Plus, planetary observers tend to shun anything that generates heat all of the time; far better to let everything equalize in temperature and then to bag some images before the next application of heat. This leads to another point. If you are observing alone on some high-altitude spot, far from your apartment, you will need loads of batteries, and I do not know of any decent battery-powered hairdryers. The best place to observe from is a balcony well above ground level; preferably one that faces south (or north if you are in the southern hemisphere). Access to unlimited electric power is so vital when you are running a telescope for hour after hour. Contrary to popular belief, the heat from buildings is not a planetary killer. The ground itself is a source of heat and your telescope has to be mounted on some form of ground unless you have an antigravity platform! A south-facing balcony within easy access of facilities is the best kind of observatory you can have. Failing that, site the telescope outdoors and pack enough electrical cable and adaptors to reach the indoor supply. Plus, do not forget the safety issues here: damp grass and high voltage do not go well together. You need to be familiar with the local electrical plug design and bring plenty of adaptors. Is the local electricity 240 volts or 115 volts? Will your telescope work on either? What, precisely, is the pin spacing on the typical electrical power socket? In 1995 and 1998 I traveled to India with eclipse and Leonid meteor expeditions. On both occasions matchsticks were provided with which to help jam your electrical appliance into the wall socket. India is renowned for its unpredictable electrical supply and wall sockets. A friend of mine stayed in an Indian hotel room where the light switches for his room were in the next door room and vice versa! To adjust the bedroom lighting you hammered on the wall so your neighbors could switch your lights on or off, regardless of what activity they were involved with at the time!

A balcony observing platform next to your bedroom is a godsend. If it is cloudy at first you can get some sleep in the evening and then roll out of bed and just stagger a few yards to your observing station. No need to strain your back hauling your equipment about, it is already set up a few yards away, ready to be wheeled out onto the balcony. If you are worried about polar alignment, don't be. You only have to do it once, and then just align the telescope tripod on marks on the floor the next time. However, locating the pole star can be a fraught business when near the equator if you cannot see Polaris. Nevertheless, by observing which way stars drift when they are near the horizon, or on the meridian, will tell you how close you are to being polar aligned. Polar alignment is not supercritical for planetary work and even rough aligning within a few degrees of the pole will allow good images to be secured. The main disadvantage will be that planets tend to drift out of the webcam field in 5 or 10 minutes with this magnitude of misalignment.

While we are on the subject of equatorial mountings, it is worth simulating (months before you fly off) just how your telescope behaves when the polar axis is much closer to being horizontal than at home. In the worst case it might not permit adjustment to that low an elevation, or it might simply fall over! At high latitudes the center of mass of your telescope may well sit directly over the mount; an ideal situation. But at low altitudes the telescope mass may try to tip the whole mount and tripod over; maybe the whole telescope and mounting will cartwheel over the balcony railing, plunge 10 stories and kill a crowd of people beneath. OK if your mother-in-law is one of them: not OK if the local Mafia boss was underneath. All of these nightmare scenarios need consideration.

Then there is the issue of ensuring the telescope arrives safely and not as a pile of broken glass and twisted plastic. While most airline baggage handlers are not reckless people, they do throw suitcases and unmarked personal luggage around in a manner not dissimilar to a rugby player. Some of them are genetically very close to being Neanderthal throwbacks. A telescope needs to be transported in a rigid equipment case, clearly labeled with banners saying things like FRAGILE, DO NOT DROP, THIS SIDE UP, and GLASS all over the outside. Baggage handlers are not psychic; they do not know what is in a box unless there is a big label. Inside the case there should be a firm foam insert, snugly housing your optical tube assembly (OTA). A professional photography type case is ideal, i.e., a lightweight aluminium casing that can be properly locked, but a homemade housing is OK providing it is rigid, dent proof, and filled with a suitable foam insert cut to the telescope OTA shape. A Newtonian telescope can be completely disassembled for transportation if required. The heavy primary mirror can be packed in its own compact box ensuring that its weight does not result in it escaping the mirror cell in transit (this does happen, even when Newtonians are delivered from professional dealers). Modest-sized primary mirrors (under 30-cm diameter) typically weigh less than 10 kilograms and can easily fit in a small box surrounded by clothing, as padding, in a domestic suitcase. Newtonian secondary mirrors are usually glued to their holders and rarely come out. But if they are not glued, removing them in transit is a consideration. Also essential when traveling abroad is a complete set of screwdrivers, wrenches, and Allen keys for every component that might come loose or need adjusting. Most astronomers when traveling abroad have to pay an extra charge for baggage allowance as they exceed their personal limit. (I know one observer who never pays extra, he just goes with his wife and bans her from taking any spare clothes!! She stinks like a mule by the end of the holiday, but that is the price you sometimes have to pay.) Make sure you know how much you will be allowed to take and how much you will be charged per kilo of extra weight. At the time of writing, packages over 32 kilograms in weight were banned by some low-cost airlines, regardless of the traveler wishing to pay an appropriate excess fee. When you arrive at some countries with a box of expensive-looking equipment in its original crate, customs officials may grab it and claim you are importing equipment. You may have to sign complex forms and retain those forms for when you leave the country: you have been warned!

One unnecessary item of extra poundage is the telescope counterweight. Carting lumps of lead on holiday is unnecessary with a bit of forward planning. Hollow counterweights made from metal cylinders can be used as substitutes. These can be filled with sand or soil (or bodily waste) at the observing site, thus saving multi-kilograms in excess baggage.

Amateur astronomers have never lacked versatility when traveling abroad. I know of one German amateur, with an axe, who made a very efficient pseudo-Dobsonian mounting from his hotel bedside table on the 1991 Mexico eclipse. I dare say the hotel authorities were unimpressed, but he was back in Germany by the time they missed it. When staying in hot countries that do not have the highest quality sanitation, make sure you drink plenty of water (to avoid dehydration) and make sure it is bottled or boiled. Also, avoid salads, cream, ice cream, or anything that isn't fully cooked. Stomach upsets can ruin a holiday and ensure no observing is done. I have had friends who have had such serious stomach upsets that they have had to fabricate a home-made rectal bung for long bus journeys. On the return journey home one of these friends suffered serious constipation. On arriving back in England he bent over to open the eyepiece drawer in his home and the acoustic volume of the resulting flatulence woke next door's tortoise from its winter hibernation. To this day the tortoise is still receiving trauma counseling.

When observing in hot countries you may well encounter annoying biting insects that come out at night. Be especially wary of mosquitoes, some of which can carry malaria. Make sure your inoculations are up to date for the area you are visiting. A multiple insect repellent strategy may be required, as depicted in Figure 5.3.

Hot countries may also cause overheating of telescope tubes. Such a problem was encountered by Damian Peach on one foreign trip and it was necessary for him to remove the telescope tube and fancool it indoors, as shown in Figure 5.4. However, such countries do enjoy very pleasant night-time temperatures (Figure 5.5), a world apart from the winter observing conditions here in the U.K. Incidentally, Damian spent three weeks in Barbados in April 2005, and from the grounds of a rented villa on the south of the island he reported clear skies and near-perfect seeing on 19 out of 21 nights. There was far less variation between the day- and nighttime temperatures on the island than one would experience on a continent and Damian declared the island to be "truly blessed with near-perfect seeing." He estimated that an instrument of 35-cm aperture would be needed to really do justice to the conditions there, although, of course, larger apertures generally have much worse thermal problems.

Whenever I have traveled on a solar eclipse expedition or any other type of astronomical expedition I have always carried out several dry runs with the equipment I intend using on the trip. It is amazing what can be overlooked. For example, when planning for a total solar eclipse, many first-timers forget that it quickly gets very dark as totality approaches. Suddenly you cannot see your equipment! This is just one example. Before you leave for the trip you need to get all the equipment

Figure 5.3. Damian Peach armed to the teeth with mosquito deterrents, ready for a nighttime observing session in Barbados! Image: Damian Peach.

that you will take abroad (no more, no less) outside, in the dark, and see if it works. If you find that you need something extra, that you had not planned to take, make a written note. Forgetting something vital can be disastrous when you are thousands of miles from home. An equipment checklist should be started months before the trip and, when something occurs to you, just add it. Here is a start for you, from my own astro-holiday master list:

Astro-Holiday Planetary Equipment Checklist

Valid passport and plane tickets;
Telescope in quality packing case;
Fragile/Glass/This Way Up labels;
Telescope cables and all accessories;
Electrical adapters for the country;
Power extension cables and multi-socket power plug-boards;

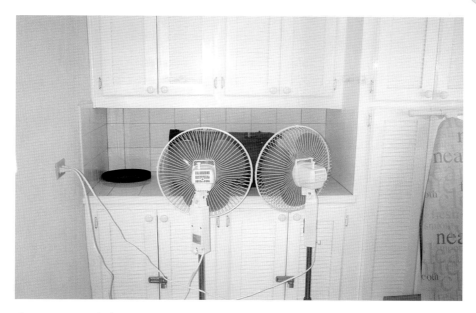

Figure 5.4. High daytime temperatures can necessitate some drastic cooling techniques on an optical tube that has become too hot. In this case, a Celestron 9.25 tube is being fan cooled indoors after getting too hot in Barbados. Image: Damian Peach.

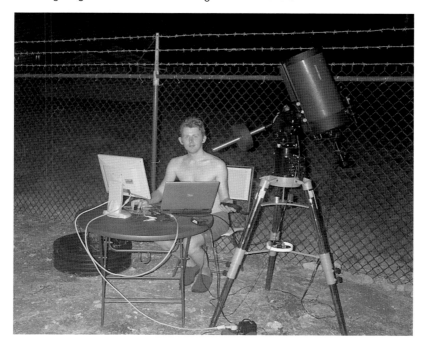

Figure 5.5. Damian Peach observing with a Celestron C9.25 while on a three-week imaging holiday in Barbados. Seeing conditions were near-perfect on most nights. Image: Damian Peach.

Astro-Holiday Planetary Equipment Checklist (Continued).

Spare batteries for the motorized focuser;

Eyepieces, Barlow lenses and *Powermates*; Hairdryer and dew heaters;

Lightweight dewcap; Ephemerides for transits of Jupiter's moons;

Internet accessibility at the hotel/apartment/Internet café;

Loads of torches and spare flashlight batteries;

Tools to collimate the telescope;

Screwdrivers and spanners for every possible adjustment/disaster;

Blank CDs for archiving the best processed images (in case of hard disk damage) *or* a portable spare hard drive/card;

Spare webcam; Filters and filter wheel (if used);

Insect repellent (the little horrors come out at night);

Sellotape, sticky labels, scissors, string, and rubber bands (they have saved my life a few times);

Red cellophane (for sticking over torches);

Wristwatch (and spare) with alarm clock and illuminated face;

Earplugs (so you can sleep in the day);

Phone numbers for reliable local amateurs or friends back home who can ship stuff out to you in an emergency;

A mobile phone that will work in the country where you are staying.

Laptops

Obviously, the traveling planetary webcam user will want to take a laptop on the trip abroad, rather than a desktop PC. However, if you are buying a laptop for the journey (or for planetary imaging in the U.K.) I would advise paying a little bit extra for a laptop with a high-quality screen. Laptop manufacturers live in a highly competitive world. Every customer is looking for more features at a lower price and if manufacturers can save a few cents here and there by using a cheap component, they will. (Trust me on this; I used to work in the electronics industry!) One of the most overlooked features of a laptop PC is the screen quality and yet it is especially crucial to achieving good focusing when imaging planets. Go for the sharpest, highest resolution, highest specification screen you can afford. You will not regret it, but you will seriously regret buying a PC with a poor quality display.

There is also the question of how large a hard disk you need to have in your laptop. Planetary AVIs are large files, typically 1 or 2 Gigabytes in size. On a good night you may take a dozen AVIs and you might be at your Caribbean retreat for 14 nights. If you are a dedicated planetary imager, planning to do much of your image processing on your return to the U.K. you will need a large hard disk! Portable memory storage units can now be acquired at very low cost. If you are taking webcam AVI videos on many nights you may want to transfer hundreds of gigabytes of data to one of these to free up your main hard disk. They are a very worthwhile investment.

CHAPTER SIX

Planetary Webcams and Their Alternatives

Imaging technology is a very fast moving area in the early 21st century. Once millions of dollars are invested into any form of technology, fueled by public consumption of a product, remarkable advances can occur. In 1989 the very first amateur CCD cameras started to appear on the market. A decade later, the first decent digital cameras and webcams started to be mass produced. In 2004/2005, Canon's Digital Rebel/300D Digital SLR put a real SLR camera in consumers' hands at a very affordable price. I mention all of this because it is likely that even in the few months it takes to print this book, more advances will be made. However, I think it very likely that a webcam-type product will still dominate planetary imaging for many years to come. It is all a question of how good and affordable that product will be. While the standard ToUcam Pro imager and its equivalents (with the same CCD chip) are fine for planetary imaging in 2005, manufacturers' product lines change with time and it is always good to be ahead of the game. Many of the world's most advanced planetary observers use custom astronomy webcams to get the best results, either so that they can use LRGB techniques with a sensitive monochrome camera, or so they can use multi-second exposures for IR and UV work. Some of these custom cameras can also be used for low-cost deep sky imaging too, killing two birds with one stone.

The prime consideration in planetary imaging is to have a sensitive, fast-download detector at the heart of the system and as little noise in the detector electronics as possible. The Sony CCDs in the best webcams already have chips with good quantum efficiency (QE) and low noise, but 100% QE must be the eventual aim. Once this is achieved it is important to reduce the readout noise and thermal noise as far as possible. Cooling the detector is used to achieve this latter requirement in long exposures, but it seems unlikely that mass-produced commercial webcams will go down this route, especially as readout, not thermal noise, dominates in short exposures. However, there are already air-cooled astronomy webcams on the market, although, frankly, the cooling does not make much difference for planetary imaging; indeed, I know amateurs who have cut the fan leads and sealed the back to reduce dust ingress. Download speed is another area where improvements can be made. Remember the Planetary Imager's Mission Statement: "To record as many high quality frames before the planet has rotated more than the resolution of the telescope." In current USB 1.1 webcams, minimal image compression takes place at 5 and 10 frames per second and exposures of 1/10th of a second are needed to get good signal-to-noise images, especially with Saturn. So, with exposures of 1/10th or 1/5th of a second being needed, frame rates higher than 10 frames per second are unachievable. The current webcams do fine. Another issue here is hard drive usage and address ability. With USB 2.0 webcams, far more data can be transferred from webcam to hard disk, but over several minutes an entire hard disk can be eaten up! Also, the frame stacking software Registax can (in 2005) only address a maximum of about 10,000 frames or 2 Gigabytes at any one time (although multiple AVIs can now be opened in parallel). Older PCs (such as used in the observatory) have operating systems that will only allow the addressing of half this amount of memory, so your webcam AVIs may have to be restricted to 2,000 frames and processed separately once you start collecting multi Gigabytes of data. These are vital considerations that need to be considered before upgrading to a USB 2.0 or Firewire download camera. Would you ever need such a system at all? Well, the speed of USB 2.0 does allow a completely uncompressed deeper bit-depth (luminance and color) transfer of data and may well be useful on very bright targets (like Venus in the UV, the Moon, Mars, or the Sun) where freezing the seeing with ultrashort exposures is of great use. But, when stacking hundreds or thousands of frames, USB 1.1 webcams do a great job with a basic PC and at an affordable price. At the time of writing, the Philips ToUcam Pro, despite being an off-the-shelf $100 webcam, designed for the public, not astronomers, is still the best color planetary imaging device and it is affordable. With a telescope adapter, UV-IR blocking filter, and Registax software you are in business!

Sony CCD chips lie at the heart of the most light-sensitive webcams. These are technically known as Interline CCDs, which means they are of the type used widely in domestic video devices and feature circuitry on the chip surface that download the charge from the pixels and to the A/D converter. The presence of electronics on the top of the photosites used to mean that Interline devices were far less sensitive: the pipe work was blocking photons. However, in recent years, Sony has perfected the use of microlenses above the pixel surface. These microlenses collect the light before the "piping" intervenes. Thus, the current generation of Interline chips are both sensitive and low-priced and therefore ideal for webcams and cheap astro-cams. The alternative Progressive Scan/Frame Transfer devices such as the Kodak KAF chips used in SBIG's cameras are far more expensive. The Sony chips that are most widely used in astronomer's webcams have the prefix ICX. There are a number of different subtypes, namely, the ICX098 series (suffixed with AK, BQ, AL, and BL) and the ICX424 series (suffixed with AQ and AL). Other Sony chips in the ICX series have been used by astro-enthusiasts to build their own modified webcams, too, namely those suffixed 254, 255, and 414. One astro-webcam manufacturer, Astro-Meccanica, uses the ICX 409AL chip in its KC381 planetary camera. All of the L suffix cameras are monochrome units with unfiltered pixels. The sensitivity of these monochrome CCDs (i.e., the voltage produced for a specific illumination) is, not surprisingly, much better than the filtered CCDs of the same type (typically, about 2.5× times better). But, of course, once R, G, and B filters are used it is back to square one (almost).

A summary of Sony's "planetary-friendly" CCD chips, i.e., those most used by amateur astronomers and astro-webcam manufacturers, is shown in Table 6.1.

While mentioning CCD webcams, I should, perhaps, mention Meade's LPI Imager/Autostar Suite, which appeared in 2003 and was marketed as a lunar and planetary imager and autoguider combined with a planetarium and telescope control package. While the package was incredible value for money, the CMOS detector in the LPI imager was noisy and insensitive compared to a ToUcam Pro and the download time was sluggish when compared to any webcam. No serious planetary imagers that I know use this detector. However, as a cheap and cheerful bundled package for complete imaging beginners, it worked well.

Astro-Webcam Manufacturers

In the last few years a number of custom companies have sprung up in response to the need for affordable CCD imaging for those amateur astronomers with a tight budget. In many cases the products offered have simply been modified webcams, altered by a few wiring changes such that long, if noisy, exposures are possible, making webcams suitable for deep sky imaging. By stacking hundreds of frames, the noise has been reduced to acceptable levels for exposures of 30 seconds or so (often the unguided technical limit of the telescope's drive). Other webcam-modified products have included air-cooled and Peltier-cooled webcams, to

Table 6.1. Summary of Sony's "planetary-friendly" CCD chips.

Chip Des.	Color	Pixels	Pix. Size (microns)	Sen[*]	Webcam Examples
ICX098BQ	Yes	640 × 480	5.6 × 5.6	100%	Philips ToUcam Pro Logitech QC Pro 3000/4000 ATiK-1C/Celestron NexImage
ICX098BL	Mono	640 × 480	5.6 × 5.6	250%	ATiK–1HS II
ICX098AK	Yes	640 × 480	5.6 × 5.6	75%	Vesta Pro/Vesta 675
ICX424AQ	Yes	640 × 480	7.4 × 7.4	75%	ATiK–2C/Lumenera 075 Colour
ICX424AL	Mono	640 × 480	7.4 × 7.4	180%	ATiK–2HS/Lumenera 075 B&W
ICX254AL	Mono	500 × 480	9.6 × 7.5	310%	SAC–8.5
ICX409AL	Mono	740 × 570	6.5 × 6.3	370%	Astromeccanica KC381

[*] The "Sen" rating is calculated by the author taking into account the parameters from the Sony data sheets. It is an approximate indicator of the relative sensitivity of the CCD chips with a planet covering the same number of pixels. The popular ToUcam Pro is rated as 100% as the standard. Not surprisingly, the monochrome CCDs are the most sensitive as they are unfiltered. In practice, these CCDs will be filtered by the user to obtain an RGB or LRGB image. The monochrome ICX254AL and ICX409AL detectors are the most sensitive CCDs used in astro-webcams but have the disadvantage of having nonsquare pixels and so need resampling by the users software.

reduce the long exposure thermal noise problems. A few of these products are of interest to the planetary imager, too, especially where a sensitive monochrome CCD chip has replaced the commercial color detector. Celestron's NexImage (Figure 6.1) has proved an attractive package because, although it is only a ToUcam Pro electronically, it comes as a bundle with a telescope compatible 31.7-mm nosepiece and with Registax software. It is not difficult to acquire adaptors or the Registax freeware but many consumers, especially novices, seem attracted by an all-in-one planetary imaging package, ready to go. But what are the alternatives to the off-the-shelf and cheap ToUcam Pro, Logitech QC Pro 3000/4000, and NexImage webcam devices?

Perhaps the most attractive unit here is ATiK Instruments' ATK-1HS II camera (Figure 6.2). Essentially it is a monochrome ToUcam Pro, with the Sony ICX098BQ CCD chip replaced with an ICX098BL and fan cooling added. The extra sensitivity gained with the monochrome system might seem pointless at first. After all, if you have to filter the monochrome detector to recover a color image then surely you are back to square one? Well, not quite. Remember the world's top imagers, like Damian Peach, use LRGB, not RGB. The colors in the image come from a filtered result, but the luminance can be unfiltered or in the passband that the planet has the most contrast/sharpness (e.g., red for Mars). Also, remember that unfiltered CCDs are very sensitive in the far red end of the spectrum and seeing is better there, too. As another consideration, bear in mind that the Bayer Matrix (Figure 6.3) and YUV decoding/compression system used in color webcams results in a very noisy

Figure 6.1. Celestron's NexImage imager is, essentially a sensitive webcam with a telescope adaptor and all the software provided. Image: Celestron.

Figure 6.2. The author's monochrome ATiK 1HS super sensitive webcam. A low-cost way of taking planetary (USB) and Deep Sky (Parallel port) images. Image: Martin Mobberley.

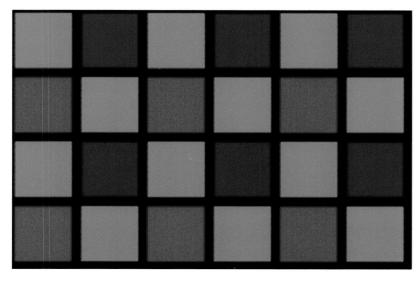

Figure 6.3. A Bayer Matrix. The pixel filters on a typical color CCD chip are arranged as shown. Every two-by-two block contains one red filter, two green filters and one blue filter. The onboard webcam processing chip on commercial webcams increasingly compresses the color (and luminance) data from this matrix as the frame rate increases. In addition, the YUV video coding system degrades the blue signal.

blue image and lower color resolution than luminance resolution. Finally, monochrome webcams can easily be used (and are often supplied) in RAW mode (see Chapter 8) in which there are no compression/artificial sharpening algorithms and the image is far less noisy. When you have seen a noisy Saturn at f/40 in a ToUcam Pro webcam and then in an ATiK-1HS II you will see what I mean. The monochrome image is far smoother. A proper RGB image taken through color filters contains far more data, too, 24 bits (3 colors × 8) or even 30 bits (3 colors × 10) and with no data compression, especially with USB 2.0. Having said that, taking a tri-color image is far more hassle, especially in the case of Jupiter where the imaging window lasts two or three minutes (see Chapter 13). However, monochrome webcams have other advantages too. Scientifically important results can be obtained by imaging planets at the extreme ends of the spectrum. For example, Jupiter can be imaged in the Methane bands at 619, 727, or 890 nanometers; the 890 band is where methane gas absorbs light the most and is well into the infrared, providing a real challenge even for red-sensitive CCDs. Venus' cloud belts reveal details at the other end of the spectrum, i.e, in the ultraviolet. Only one filter at a time is required (not three as in RGB) for this type of work and a monochrome detector is therefore ideal. The color filters covering every 2 × 2 pixel block in the color Sony chips are red (one pixel), green (two pixels), and blue (one pixel). Thus, every pixel is filtered in some way. Adding another narrow band filter at the end of the spectrum will really hammer the response of a color webcam and a monochrome camera will perform far better for scientific narrow-band work.

One of the beauties of a webcam imager like the ToUcam is its light weight. Even the flimsiest telescopic set-up does not need strengthening or rebalancing when a ToUcam webcam is added. However, when a filter wheel is also added and filters are selected during imaging, a lightweight system may be compromised, especially when the field of view is only one arc-minute; even the slightest force applied to a telescope on a flimsy mount, especially a long Newtonian, can push the planet out of the field.

The ATiK system offers a lightweight and inexpensive filter holder that can be inserted between ATiK camera and telescope drawtube. Obviously any additional light path travel between Barlow/Powermate and CCD chip increases the effective f-ratio, so the thinner the filter holder/wheel the better. A slimline, lightweight, motorized filter wheel is, perhaps, the best system.

An extra bonus of the ATiK system is that, by connecting the parallel port supplied with the camera, single-shot exposures of several seconds can be employed too, for example, when imaging through narrow-band filters or recording faint planetary moons. (If you are buying a modern laptop, check it still has a parallel port if you have a parallel port camera; many modern laptops do not and a USB port replicator may be needed.)

ATiK offer four different "modified webcam" type cameras. In addition to the ATK-1HS II, which I think of as a monochrome fan-cooled ToUcam, they also do the 5.6 micron pixel ATK-1C II. This is essentially just an air-cooled color ToUcam Pro with a telescope adapter and a long exposure capability. The remaining two ATiK offerings feature 7.4 micron pixel versions of the above cameras, designated the ATK-2HS (mono) and ATK-2C (color). These type 424 chips cover almost 75% more area than the 098 detectors and have a physical imaging area of 4.7 mm × 3.6 mm, which could be a consideration for the deep sky imager. But the CCDs are no more sensitive in use than the 5.6 micron pixel ATiK devices.

Another couple of interesting Sony CCD-based USB cameras, this time made by a Florida-based company, are the SAC 8.5 and SAC II. The SAC 8.5 uses the ultra sensitive ICX254AL EX-VIEW HAD monochrome Sony chip, which is about 20% more sensitive than even the ICX098BL. However, the ICX254AL does have non square pixels, which may be off-putting for some. Resampling the pixels to produce the correct aspect ratio is a formality with modern software but many will still prefer square pixels so they have the same resolution sampling in horizontal and vertical axes. Like the ATiK 1-HS II camera range, the SAC 8.5 has a filter holder for RGB, LRGB, CYM, and LCYM imaging, although in the case of the SAC unit an actual filter wheel is involved, so there is less chance of losing the planet when you insert the new filter in place, and less chance of dropping the filter. As with ATiK's devices, the SAC 8.5 can be used for planetary imaging with rapid AVI download or for deep sky imaging, and the price of under $600 is a lot less than the price of the non-interline, Kodak chip-based cameras. The SAC 8.5 is fairly unusual as it features Peltier, not air cooling, in this economic price range. An even cheaper unit, specifically for planetary imaging, is the SAC II. At the time of writing, this new camera was just being introduced and featured 24-bit color imaging at the remarkable price of $149.

Video Options

Webcams are not unique in their ability to freeze the atmospheric turbulence. For many years before the webcam, low-light video and security cameras where used to freeze the seeing, with the added advantage that data can be stored on video-tape. In 2005, this data can be stored on a variety of tape formats and on DVD as well. Indeed, the two technologies are now merging, as domestic video recorders featuring terabyte hard disks are now on the market. This book is about lunar and planetary webcam imaging, but I think I would be wrong to leave out video options altogether. One never knows when the technology trend will swing back the other way and some observers are far happier using video or camcorder type equipment, especially on bright objects or when timing events like an asteroid occultation where a time marker is easily superimposed on the recording.

One video option for the planetary imager is the KC381 camera produced by the Italian company Astromeccanica, run by the keen planetary imager Paolo Lazzarotti. Using the most sensitive interline CCD in the Sony ICX range, the ICX409AL, the KC381 (Figure 6.4) is a potent camera, but the pixels are slightly nonsquare at 6.5×6.3 microns. Again, while this is easily corrected in software, some astro-imagers are put off by sampling with nonsquare pixels. The KC381 is a monochrome camera so color filters will be required to construct a color image.

One videocamera that has been proven to perform on bright astronomical targets is the Neptune 100, manufactured by the Japanese company Watec. Priced at around $650, the Neptune 100 is a videocamera sensitive to 0.001 lux and with a spectral sensitivity from the visual band through to the near infrared at 940 nm. For solar imaging, or imaging of Mercury, Venus, and the Moon, it is a good choice for those who prefer to save 1/60th second video frames rather than fill their hard disk. Of course, a set of color filters will be needed too.

Another option is the Lumenera Lu070 (pcb) and Lu075 (enclosed) range of USB 2.0 VGA cameras (Figure 6.5). Lumenera is a company based in Ottawa,

Figure 6.4. Astromeccanica's KC381 CCD camera. Image: courtesy Paolo Lazzarotti.

Ontario, in Canada. Their 070/075 range currently comprises four models (color, monochrome, and enclosed or modular versions of each) that can convey uncompressed, 8 or 10 bit, 640 × 480 pixel frames at 60 frames per second down a 480 Mbit/second USB 2.0 line. (Say goodbye to your hard disk space!) No frame grabber is required with this system. The video images just come straight down the USB cable. Again, like the Neptune 100, this is an especially excellent option for bright objects where exposures of 1/60th second are long enough. The Lumenera

Figure 6.5. Lumenera's USB 2.0 fast download LU 075 camera. Fast frame rates without compression are possible. Image: courtesy Paolo Lazzarotti.

cameras store their video images onto hard disk using the SEQ video file format. Software can easily convert this into AVI format (for Registax) at a rate of about 1000 frames per minute. According to tests carried out by Damian Peach, there is little difference between the 8- and 10-bit imaging modes, and the SEQ 8-bit files are about 25–30% smaller than an equivalent AVI file. A great feature that the Lumenera Streampix software has is that you can pause the AVI capture at any time, meaning that if a cloud passes over the object you can pause the video recording. In tests carried out by Damian using a monochrome Lumenera LU075 camera while observing Jupiter from Barbados, he concluded that it easily outperformed even his (USB 1.1) ATiK 1HS. The Lu075 could easily work at 18 to 34 frames per second and give acceptable, smooth results, where the ATiK was happier at 5 or 10 frames per second. One thing to bear in mind is that the Lumenera does have 7.4 micron pixels (c.f. 5.6 micron in the ToUcam/ATiK) so focal lengths have to be increased by 32% to deliver the same sampling resolution with respect to the smaller pixel webcams. In addition, please note that these excellent cameras start at just under $1,000, so are considerably more expensive than webcams or modified webcams.

Adirondack Video Astronomy (AVA) is one of the best and most innovative astronomy dealers in the U.S., and it is not surprising that they stock a comprehensive range of the latest CCD and webcam-based products (such as the ATiK range). Their website at **http://www.astrovid.com** is well worth checking out whether you live in the U.S. or elsewhere. AVA offer a couple of in-house products called Astrovid Color Planetcams that can be used for planetary imaging. While these tiny (190 gram) video cameras do not offer huge benefits over their rivals, they do have a number of features that may appeal to the beginner. Firstly, they feature small, 4.2 micron, pixels. For a beginner, video-filming the Moon at, say, 0.4 arc-seconds per pixel, a focal length of only two meters will be required. So a 20-cm f/10 SCT will fit the bill without any Barlow lens or Powermate being required. A simple 2 × Barlow will give all the resolution a beginner would want even on a night of steady seeing. These cameras also feature a rather unusual negative image feature enabling the planet to be viewed on the monitor as a "live" color negative when trying to bring out details in bright regions. The advanced model of the Astrovid Color PlanetCam, titled the "Computer Controlled EEPROM" version, can have its settings (gain, gamma, sharpening, color balance, exposure) controlled remotely from a PC and the settings can be stored for each planet imaged. This is a useful feature e.g., for live planetarium/observatory imaging of the planets. However, despite these unique features, the planetary imaging enthusiast will probably greatly prefer the webcam approach, as stacking hundreds of images to create the smoothest image is his or her aim and not live video imaging sessions in front of an audience.

CHAPTER SEVEN

A Beginner's Guide to Using a Webcam

Many readers of this book will be complete webcam beginners. If this describes you, this is the best chapter to read before you embark on your journey. As we have seen, in the 21st century, all you need to capture stunning images of the planets is a PC with a USB port, a webcam, an adaptor, and a 20-cm, or larger, telescope. Needless to say, the telescope should have good optics and the optics should be precisely collimated. I have covered collimation in Chapter 3.

Webcams, and their associated software are, without a doubt, the best thing since sliced bread, at least from the amateur astronomer's viewpoint. Why? Because even when the Earth's atmosphere makes a planet look like it's under a jug of boiling water, if you can collect thousands of frames, and only 10% are reasonably sharp, you can still salvage a better image than the professionals were getting before Hubble was launched!

Find the USB Port and Buy a Webcam

The first step on your webcam journey is to check that you have a PC with a USB port. If your computer was made before 1998 it may not have one, in which case there is not much point proceeding. A USB port looks like a small slot in the PC and is completely different in appearance from the multi-pin serial and parallel ports. In addition, if your PC's operating system is earlier than Windows '98 then USB will not be supported. PCs and operating systems prior to 1998 simply will not work with USB webcams. Do not worry if the PC does not have high-speed USB 2.0; the basic USB is fine.

If you do have a modern PC with a USB port, great, you can proceed with buying a webcam. The best cheap webcam for lunar and planetary imaging, at the time of writing, is the Philips ToUcam Pro (Figure 7.1). It is very sensitive to light (the sensor is a true CCD, not a CMOS chip) and that is why it is so good. The 2005 model is called the Philips ToUcam Pro II PCVC 840K and will work with Microsoft Windows 98, 2000, ME, and XP. Of course, technology moves on all of the time and by the time this book is published the ToUcam Pro may not be the best on the market. It is a good idea to check on the Internet and see what the top planetary imagers appear to be using at any one time. Proper CCD-based webcams had far superior performance to noisier CMOS versions when this book went to press.

I ordered both my ToUcam Pros online, from Amazon. At the time of writing, the ToUcam Pro II was priced at around $100. Electronically, the original, white, egg-shaped ToUcam Pro I and the new, silver, ToUcam Pro II are identical.

The webcam, as purchased, will not fit onto your telescope, so you also need to purchase an adaptor. One end of this adaptor must be about 31.5 mm in diameter (to fit into the 31.7-mm [1.25-inch] diameter hole of astronomical telescopes and accessories): the other end needs to be much smaller and threaded to fit the tiny thread which the supplied webcam lens screws into. Fortunately there are many suppliers of 31.7 mm to webcam thread adaptors, many of whom advertise in the popular astronomy magazines like *Sky & Telescope*. In fact, the ToUcam Pro is so popular that you can buy one with an adaptor included, as a complete package, from many astronomy suppliers. A third (optional) accessory is a so-called UV-IR

Figure 7.1. The Philips ToUcam Pro II with the supplied lens removed. The tiny CCD chip can just be seen. Image: Martin Mobberley.

blocking filter to give a more natural color response, but this is certainly not essential for first-time users. The Celestron NexImage planetary imager is simply a ToUcam Pro's electronics in a moulded telescope-compatible casing, complete with bundled software.

The other thing to think about at this stage is how to enlarge the image of a planet falling onto the CCD chip. A good starting point is to increase the telescope's focal ratio until it is at least f/20, in other words, the focal length is 20 times the telescope's aperture. With an f/10 Schmidt-Cassegrain you will need to purchase a 2 × Barlow lens. With a Newtonian of, say, f/5 or f/6, you will need something like a TeleVue Powermate; the 5× version is excellent (Figure 7.2) and, because it is in a 31.7-mm housing, is far cheaper than the 50.8-mm 4 × Powermate. Eventually you will experiment with focal ratios as high as f/30 or 40, but f/20 is a good place to start. Bear in mind that the precise f-ratio will depend on the distance over which the image is projected by the lens. A 3 × Barlow lens projecting 100 millimeters beyond the top surface of the unit will become a 5 × Barlow and a 5 × Powermate projecting 100 millimeters further than that point will almost become an 8 × Barlow.

Image Scale Considerations

Let us have a look at this image scale business in more detail because it is important to understand the issues involved. Most of the popular webcams have CCD chips with an array of 640 × 480 pixels. These pixels are, typically, 5.6 microns in size. So the imaging chips themselves are roughly 3.6 × 2.7 mm across. The standard formula for calculating the resolution of a telescope, in arc-seconds, is given either by 138/D (the Rayleigh limit) or 116/D (the Dawes limit), where D is the

Figure 7.2. A TeleVue 5x Powermate attached to a ToUcam Pro webcam ready for insertion into the telescope drawtube. Image: Martin Mobberley.

aperture in millimeters. The first formula is based on the theoretical effects of diffraction at the wavelength of green light, whereas the second is based on practical measurements (by a chap called the Rev. Dawes) on the ability of refractors to visually split close double stars. I will adopt the 116/D formula here. Essentially, the Dawes limit tells us that a telescope of 116-mm aperture will split two double stars of equal brightness that are 1 arc-second apart. A telescope of double the aperture, i.e., 232 mm, will split two stars 0.5 arc-seconds apart. The question is, if we want to capture all the detail that the telescope can theoretically resolve, what focal length do we need to increase the telescope to? This is not as easy a question as it might sound. While the Dawes limit is established in astronomical literature, amateurs have always been able to see much smaller, high-contrast features on the Moon (like lunar rilles) and even on the planets (like the Encke division on Saturn). Resolving two equally bright components of a double star, as their diffraction patterns merge, is a special case and it tells us little about how many pixels should straddle the area in order to capture every last bit of detail. One guideline, often quoted, is the Nyquist sampling theorem, which tells us that to accurately sample a periodic signal of highest frequency f, we need to sample it at twice the frequency. From the terminology the reader will gather that this theorem is generally intended for sampling signals in the world of electronic communications. However, taken at face value it says that if a telescope can resolve 1 arc-second, we should sample the image at a scale of two pixels per arc-second. But pixels are two dimensional and across their diagonal they are 1.414 times bigger, so the diagonal sampling could be interpreted as 1.414 times too coarse. Maybe we need to sample at nearer three pixels per arc-second for a 116mm aperture telescope? With such a telescope and a webcam having 5.6 micron pixels, sampling at two and three arc-seconds per pixel corresponds to f-ratios of 20 and 30, pretty close to what many leading amateurs actually use. Beginners tend to start at f/20: more experienced users tend to drift toward f/30 or more to capture the finest details. Even f/40 is justified under good seeing conditions and with a sensitive webcam.

It should be remembered in this context that not only do planets consist of a myriad of overlapping diffraction patterns from each point on the planet's surface but the webcam user is going to end up stacking hundreds or thousands of images on top of one another. The first point is the reason why we can glimpse such fine high-contrast detail on tiny features like the Moon's lunar rilles, and the second gives us an additional statistical advantage that even the most eagle-eyed visual observer simply does not have. When hundreds of webcam frames are stacked and image processed, quite astonishing details emerge; features like Saturn's Encke division can become almost routine targets given good enough seeing. So, bearing all this in mind does anyone know just how high an f-ratio the planetary imager should really use, in perfect conditions? Is even f/30 enough? Well, in 2004, Damian Peach carried out some interesting experiments with an 80-mm aperture Vixen apochromat of exceptional quality. He imaged high-contrast features like the Moon and sunspots to try to answer this very question. Using only an 80-mm aperture, often under superb seeing conditions (even by large aperture standards), Damian could be sure that the Earth's atmosphere was having no effect on the images he was taking. The lunar pictures Damian obtained under these conditions, when he was temporarily without a decent aperture instrument, were quite an eye opener. To digress for a moment, many years ago I knew the legendary lunar and planetary photographer Horace Dall, inventor of the Dall-Kirkham Cassegrain telescope. He was the best amateur lunar and planetary photographer of his generation and he used a 39-cm aperture Dall-Kirkham telescope. I have compared Dall's best lunar pictures, taken in the 1960s and 70s, with Damian's best webcam images using his 80-mm apochromat. Damian's images resolve fractionally more than Dall's, despite the fact that Dall's telescope had nearly 5 times the theoretical resolution and over 20 times the light grasp! Features like the Hadley rille on the Moon are clearly detectable on Damian's pictures, despite being no more than an arc-second across and the theoretical resolution of the telescope being only 1.45 arc-seconds. Damian concluded that to squeeze the last drop out of the telescope's resolution he had to increase the apochromat's f-ratio to around 45. With the AtiK webcam he was using (also with 5.6 micron pixels) this corresponded to a sampling resolution of 4.5 times finer than the theoretical resolution of the telescope.

With larger apertures, Damian and others tend to use f-ratios of 30-something when the seeing conditions more determine the final outcome with a decent aperture. There are other factors, too, when imaging the planets. Saturn, in particular, is a faint world at high f-ratios. This is hardly surprising as it orbits the Sun at a mean distance of 1400 million kilometers where sunlight is only 1/90th as powerful as here on the Earth. Using a commercial color webcam at f/40 may cause the color balance to fail as the light levels are low. Also, don't forget how small the webcam field of view will become at long f-ratios. Jupiter will fill a webcam screen completely at focal lengths above about 10 meters.

A useful formula that tells you the image scale of your system in arc-seconds per pixel is as follows: image scale = 206 × pixel size in microns/focal length in millimeters. So, as an example, for a 250-mm aperture telescope working at f/30 with a webcam having 5.6 micron pixels: image Scale = 206 × 5.6/(30 × 250) = 0.15 arc-seconds per pixel. Most of the world's top planetary imagers are now working at images scales of between 0.1 and 0.2 arc-seconds per pixel.

Barlows and Powermates

Most of the world's top imagers use Barlow or Powermate lenses to stretch their telescope focal lengths to the appropriate f/20, 30, or 40 focal ratios. Because of the renowned quality of the devices made by the U.S.'s TeleVue Corporation, founded by Al Nagler, most leading amateurs use their systems for this purpose. In fact, the trade name Powermate is reserved by TeleVue. Any optical system in the telescope light path can degrade the image, so it is a good idea to use a quality component, especially at short telescope f-ratios where aberrations are more common. TeleVue's Barlow lenses and Powermates come in a variety of f-ratio enlarging values. At the time of writing, 2 × and 3 × Barlows were available as well as 2 ×, 2.5 ×, 4 ×, and 5 × Powermates. For narrow-field planetary work there is little difference in the performance of the 2 × Barlow and the 2 × Powermate, however, the Powermates dominate at the high-power (4 and 5 ×) range. The 5 × Powermate is especially popular with Newtonian users who wish to extend their relatively modest f-ratios. In essence, all a basic Barlow lens needs to be is a negative, concave (diverging) lens, like the sort you would find in a pair of short-sighted user's spectacles. (Perhaps confusingly, when used in that application, a negative lens actually makes objects look smaller!) However, in practice a well-made Barlow lens will not add any color aberration to the image and will be multi-coated for maximum light transmission. The Powermates go one step further by featuring an additional field lens to reduce vignetting and allow much higher f-ratio enlargement.

Confusion often arises when using Barlows and Powermates as, on inspection, you often find that the f-ratio you get is a bit larger than you expected. Fortunately, TeleVue have some very useful diagrams on their web pages that show how the enlargement factor for all their lenses increases with the projection distance from lens to CCD chip. When you put accessories like filter wheels in the light path, between Barlow/Powermate and the webcam chip it is surprising just how much the f-ratio can increase beyond the nominal 2, 3, 4, or 5 ×. There is another source of focal length enlargement, too. When a compound instrument like a Schmidt-Cassegrain is employed, if the instrument is focused by moving the primary or secondary mirrors, the focal length of the instrument varies. Predicting exactly what f-ratio your system will end up being can be an almost impossible calculation, but if you assume it will be some 20 to 30% more than you would expect, you will not be surprised. For example, I have a 250-mm f/6.3 Newtonian and a 5 × Powermate, so I would expect an f-ratio of around 31.5. In practice, with no intervening filters, I get f/38, a 20% increase. Knowing this can actually save you some money. For example, you may feel that your system requires a 4 × Powermate, which, at the time of writing, is only available in a large (50.8-mm eyepiece diameter) barrel size. However, if you take a 30% enlargement into account, you may find that a 3 × Barlow will suffice, as it may give nearly 4 × magnification in actual use. At current prices that will save you 60%. In extreme cases, a 3 × Barlow projecting 100 mm beyond the end of the tube will give 5 × enlargement. For a 5 × Powermate, the extra projection distance yields a whopping 8× magnification.

Getting Familiar with a Webcam

If you are a complete webcam beginner, once your webcam arrives, simply follow the instructions in the box and install the supplied software. You may choose to ultimately use different software to control the camera, but you must install the proper manufacturer's drivers, as directed, to start with. In practice, you will simply be using the supplied software to record what is called an "AVI" video on the hard disk. The Philips routine that does this is called Vrecord, but it will all depend on what webcam you ordered.

If you have never used a webcam before, it is essential to spend a few days playing with it and the manufacturer's software indoors with the webcam's normal lens attached. Get to know how to use the webcam in "manual" mode before you try connecting it to the telescope, i.e., untick the "auto" box (see diagrams). In passing, it is worth mentioning that the tiny webcam lenses, supplied with commercial units, usually have a crude infrared blocking filter, in the form of a coating or a plastic strip, attached to the rear lens. Once this lens is removed, the full sensitivity of the CCD chip, which reaches much further into the infrared (and a bit further into the UV) than the human eye will be revealed. While more sensitivity to light may seem a desirable feature, it does mean that when planets are low down the color dispersion will be considerable. Thus, many amateurs purchase an inexpensive UV-IR blocking filter to screw into their webcam adaptor or Barlow lens to restrict the spectral range and give truer colors, too. But for first steps, such a filter is not necessary.

Once you have got to grips with your webcam software and know how to save an AVI video to disk, check just how much of your hard disk is disappearing. At 640 × 480 resolution, each color frame of a TOUcam video takes up almost half a megabyte. You can eat your hard disk up at a rate of 4 or 5 megabytes a second when storing AVI videos at 10 frames per second: you have been warned! A frame rate of 10 frames per second is a good one to choose. Faster frame rates tend to corrupt the finest detail due to image compression.

Alternatives to the manufacturer's webcam software exist aplenty. The webcam-friendly astro-package, called IRIS, which has a webcam viewing/recording option can be downloaded from http://www.astrosurf.com/buil/us/iris/iris.htm.

The K3CCD Tools package can also be used. K3CCDTools can be downloaded from http://www.pk3.org/Astro. This software, by Peter Katreniak, can also be used to align and stack webcam AVI frames in much the same way as Registax, although Registax is far more widely known and, in my view, a far more versatile package.

At the Telescope

Now that you are familiar with the webcam it is time to join it to the telescope and really start imaging. I will offer an extra word of advice here. Once you unscrew the webcam lens, the chip inside is exposed. You certainly do not want to let extra dust into the lens hole and on to the chip, so *never* leave that hole where the lens used to go unplugged. I leave my webcam permanently attached to the Barlow lens, but if you do not want to do that, a plastic cap that fits over the webcam adapter is a good solution. If you do get dust on the chip, a computer duster dry air aerosol spray can be used to blow it off.

The USB cable that comes with a few webcams can be a miserly 1.5 meters long. However, USB should work reliably up to 5 meters or more, so buying an extension cable should allow a bit more flexibility. Note: you need a USB A (male) to USB A (female) extension lead. The world's leading planetary imagers do not observe remotely, from indoors, as fine positioning of a planet on a chip, to sub arc-minute precision, is best done with the observer outside, at the telescope. Initially assessing the seeing conditions is best done visually, too. The PC usually needs to be outside, unless the telescope is sited suitably close to a patio window through which the USB cable can pass, and you have a motorized focuser and a long telescope handset cable. A laptop is a good option, too, as lugging PC hardware outside is a tedious business. Critical focusing is vitally important in webcam work, so you need to be able to see the PC screen as you focus in real time. In other words, there must be a good feedback loop from the observer's eyes watching the PC screen to his or her focusing hand. Undoubtedly, the Moon is the best target to aim for when you try your first nighttime trials.

Here is a step-by-step nighttime schedule, based on my own outdoor operations:

1. Make sure the webcam is plugged into the USB port before the PC is switched on.
2. Then power up the PC and run your webcam software (e.g., IRIS or K3CCDTools).

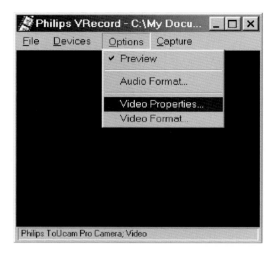

Figure 7.3. Selecting the Video Properties in the Phillips driver software.

3. Once the webcam is on, check it is set to 640 × 480 resolution by accessing the Video Properties box via the manufacturer's or third party software (see Figures 7.3 and 7.4).

4. Call up the manufacturer's webcam image and camera control windows and disable the auto settings; we want full manual control of exposure and frame rate.

5. Turn the gain up close to maximum and select an initial frame rate of about 10 frames per second and 1/25th second exposure. (In manual mode

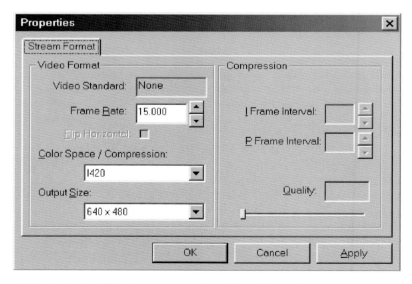

Figure 7.4. Setting the Video Properties to allow 640 × 480 resolution.

this 1/25th value is meaningless! The exposure time is 1/10th second at 10 frames per second).

6. Connect the webcam and its adapter to the Barlow lens/Powermate.

7. Shine a torch into the Barlow lens and check on the screen that the webcam image brightens.

8. Position the telescope visually, such that a planet is exactly in the middle of the eyepiece. Actually, the Moon may be a much easier target for beginners.

9. Remove the eyepiece and put the webcam/adaptor/Barlow combination in the drawtube. A blurry blob (the out-of-focus planet or lunar surface) should be visible on the PC screen. If you have a shaky mounting you may have to be *very* careful at this stage. The extra webcam/Barlow weight may cause the tube to drop and the planet will disappear. Experience at estimating the degree of droop is invaluable here. Ultimately, a separate guide telescope, with cross hairs, may be required for perfect centering of a planet in the field. Another option is a flip-mirror finder, although this will complicate issues like the available focal length, especially with a Newtonian.

10. Focus the telescope until the target is nice and sharp. For Jupiter, you may prefer to temporarily increase the webcam gain and focus on one of the nearby Galilean moons; for Saturn, the Cassini division is a good focusing guide. Investing in a smooth motorized focuser will make things far less shaky when you touch the focuser. For the Moon, any craters near the terminator make good focusing targets.

11. Once you are satisfied with the focus (when seeing is poor, you are never satisfied), set the frame rate to, say, 5 or 10 frames per second and keep the exposure to 1/25th (see Figures 7.5 and 7.6). Then, with the brightness setting midway, alter the gain until the planet is reasonably bright, but none of it is saturating. You can now close the webcam image and control properties windows.

12. Fine position the planet away from any known blemishes on the webcam chip.

13. Call up the webcam's Capture menu and choose a time and frame rate and a filename, e.g., 120 seconds, 10 frames per second, Jupiter1. Hit return!

The software will then start saving your webcam output to the hard disk at a frightening rate. The best stacking software, called Registax, can handle file sizes of up to 2 Gigabytes, which equates to about 4,500 frames at 640 × 480 pixel resolution. Registax' absolute frame limit is 10,000 frames.

Now it's time to come indoors and tackle the image processing. A word of caution may be necessary though. We do not need any sound recorded with our webcam AVIs. It is surplus information. So, in the Philips driver Properties window, make sure the audio box is unticked before the webcam AVI is saved. Apart from saving unwanted data, having the webcam microphone on can, occasionally, lead to an "error decompressing AVI" message when Registax is initiated. In this case, your night's work will have been wasted. I always check that Registax will open my saved AVI while I am still outside at the telescope.

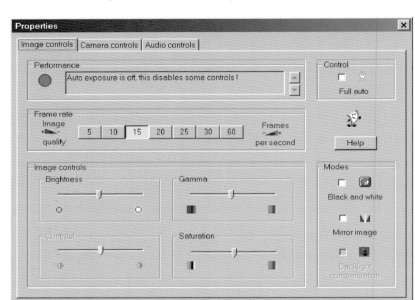

Figure 7.5. Setting the frame rate, gamma, brightness, and saturation can be achieved from the Philips Image Controls window.

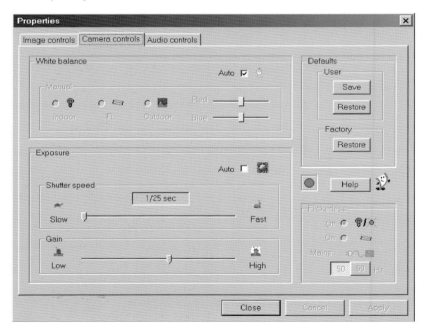

Figure 7.6. Setting the exposure time and webcam gain can be achieved from the Philips Camera Controls window.

Once you have mastered the art of saving a webcam AVI file to your hard disk, it is time to indulge in some black magic that will turn the noisy mess on your PC screen into a quality image. We have not yet mastered any of the "tricks of the trade" of planetary imaging, but more on those aspects later.

Throughout the early 1990s, when planetary CCD imaging was in its infancy, it was standard practice to take a hundred or more planetary images throughout the night, just waiting for one good sharp one to come along, courtesy of a few stable atmospheric moments. When that one good image came along, if the observer had not expired from frostbite, it was time to pack up and go inside! In the late 1990s, amateur planetary imagers started using fast download CCD cameras and collecting enough images in a few minutes to stack dozens of good images together. Then webcams entered the equation and it was suddenly possible to capture hundreds or thousands of good frames before the planet had rotated too much. When this volume of good frames is involved, some special software is needed to stack the images automatically. The package that emerged head and shoulders above the competitors was written by a Dutchman, Cor Berrevoets (Figure 7.7) and is called Registax. This package is, without doubt, the most popular AVI frame stacking software and can be downloaded from http://aberrator.astronomy.net/registax/. Registax is supplied with Celestron's NexImage Solar System Imager.

Once you have downloaded Registax from the Aberrator web page, you just run the downloaded set-up file and the program then installs itself like any other

Figure 7.7. Cor Berrevoets. The software genius who transformed the process of stacking hundreds and thousands of webcam frames and thus opened up a new era of affordable planetary imaging to amateurs worldwide. Image: Cor Berrevoets.

Windows application. For simple image stacking, Registax is a snap to learn. The opening window is shown in Figure 7.8. Help files in Word format come with the download and these are more than sufficient for mastering the basic aligning and stacking of images. Once Registax is running, you just hit "Select" and load the AVI file you want (e.g., Jupiter1.avi). Make sure color processing is ticked and then set the processing area to 512 pixels and the alignment box to 256 pixels. If you tick the "Frame List" box, you will create a window listing the hundreds or thousands of frames in the video and by clicking on a frame and using the keyboard down arrow you can look at every image in the video. You can also manually select good frames and deselect bad frames for aligning and stacking. If you do not want to wade through each frame, you can just search for a really good one and choose that as the reference shape with which to align and stack all the others. When you have found a really sharp frame, with Saturn's Cassini division nice and sharp or Jupiter's moons looking like tiny circles, just move the mouse to drag the square alignment box to surround the planet and then right click the mouse. This registers the reference image and takes you to the aligning and stacking phase. More detailed comments on Registax' advanced features are given later in this book.

There are numerous options for aligning and stacking at this point, but, for initial first-night results, the default values can be used to give a reasonable result. You can, at this stage, just initiate the aligning, optimizing, and stacking processes and the software will then run, aligning and stacking all the images that qualify for the default

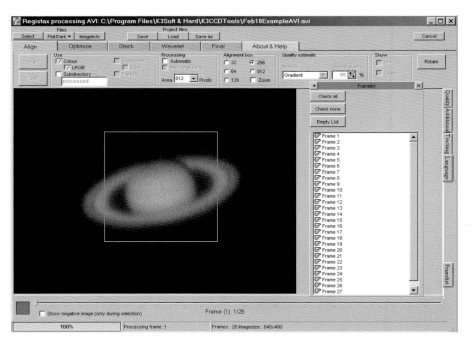

Figure 7.8. The opening window of Registax. After clicking "Select" (top left) and loading the AVI file, thousands of frames can be inspected by clicking 'Framelist' in the bottom right-hand corner. Mouse clicking on a sharp frame then registers that frame as the master shape with which to align and stack the others.

"80%" quality setting. Depending on the number of frames and the speed of your processor, the aligning and stacking process may take a few minutes or even two hours (you will wish you had a faster PC), but the end result will be a very bright, very clean stack of the best images in your AVI video. Your best bet at this point is to play it safe and click the save button, saving the image as a single bitmap (or jpeg). When you are more familiar with Registax you can learn how to do a better job of aligning and stacking only the best images. Avoiding the "Optimize" process will make the whole stacking and aligning process run much more quickly. I have found that turning "Optimize" off also sometimes eliminates fixed-pattern electrical noise, which seems to be made worse when "Optimize" is turned on with a poor set of frames.

When all the images have been stacked, the image processing window shown in Figure 7.9 appears. At this stage, Registax can provide some very powerful routines and, to get the best results, you will want to endlessly tweak the so-called wavelet layer sliders and the contrast, brightness, and gamma values. The wavelet sliders look scary but what you need to remember is that the top (layers 1 and 2) sliders enhance or suppress the finest details (like noisy pixels), whereas the bottom sliders (layers 5 and 6) enhance and suppress much larger features. They act a bit like an Unsharp Mask feature in other packages. Tweaking them endlessly will both delight and frustrate (see Figure 7.10). As a first step, sliding the layer 4 slider to the right-hand side will show you how detail can be enhanced, but I will have more to say on all this later on.

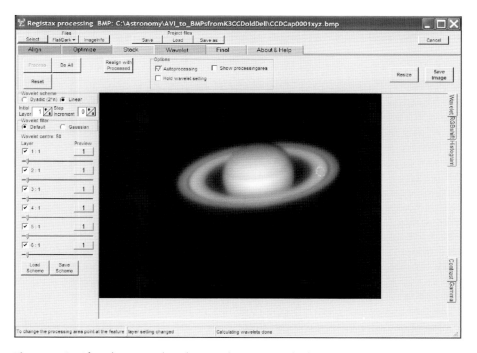

Figure 7.9. After aligning and stacking (and optimizing) the frames, Registax presents you with the "wavelet" window. The sliders (on the left hand) are initially set to one, but increasingly sharpen fine detail when moved from left to right. The top sliders sharpen the finest detail and the bottom sliders sharpen coarser detail.

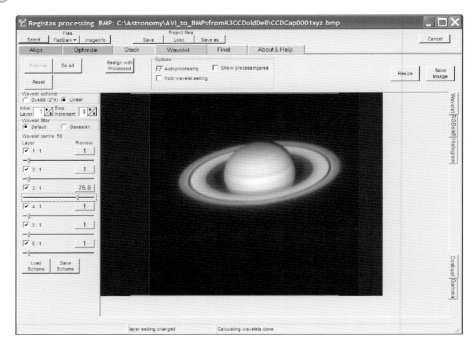

Figure 7.10. The same wavelet window as shown in Figure 7.9, but with slider 3 moved to the right. The Saturn image has become noticeably sharper.

Bitmaps, TIFFS, and JPEGS

Before we leave this introduction to webcams, I would like to add a few words on the subject of image file formats, just in case anyone embarking on webcam work is an image processing novice, too. Most images that are displayed on the web or are sent in e-mails are in the jpeg format and have a file name ending in ".jpg." Unlike large, lossless, bitmap (.bmp) and tiff (.tif) files, jpeg files feature lossy image compression. In other words, the size the image takes up on your PC can be reduced considerably at the cost of image quality. Jpeg images are fairly compact, even with zero compression, but to make them smaller they employ a clever mathematical transform that looks at each 8×8 pixel block and creates a compact code to economically summarize the data (color and brightness) within each block. With the maximum compression (lowest quality and smallest file size), some 8×8 blocks will actually dissolve into one solid 8×8 slab with no interior color or brightness variations. We do *not* want this, as it will wreck our planetary images. Planetary images are relatively small and the jpeg options should always be set to highest quality/lowest compression/largest file size to preserve all our good work. Use the jpeg format, to save your work, but always have it set to the highest quality. Registax always does this anyway, by default, but other packages, especially if someone else has used your PC before you, do not. You have been warned!

For the highest quality, save the image stack in 'FITS' 32 bit format before using the wavelet sliders. (FITS=Flexible Image Transport System).

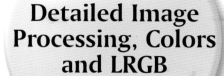

CHAPTER EIGHT

Detailed Image Processing, Colors and LRGB

Advanced Techniques

Images from the world's best planetary observers go well beyond the basic stacking and processing techniques we have discussed so far. To quote Damian Peach, getting the best results is just an endless round of "tweaking, tweaking tweaking." Beginners in planetary imaging often make the same mistake, namely they try to push the processing too far to try to compensate for poor seeing or a poorly collimated telescope. There is only so much that image processing can do and if the original raw image is poor, no amount of image processing will fool anyone. Under such circumstances, a "natural planetary appearance" is far better than an overprocessed one. What are the true planetary colors. This is a *very* difficult question to address. Different PC screens show colors differently and space probe pictures of planets have not been taken through the Earth's atmosphere, which adds a bluish tint to everything. Space probe pictures of the planets seem to show subtle variations on each mission and their images rarely look identical to Hubble images either. Professionals and amateurs all process their images to enhance detail and color, so the whole area of "true color" is a minefield. Indeed, it has been debated to death in many astro-forums and a whole book could be written on it. My best advice would be to tweak the image until it looks like the view through the eyepiece. What do the best amateurs do and how do they achieve their sharp but natural-looking color planetary images? It is very instructive in all these instances to simply try to track down precisely how they get their results: just why are they so good? Are they lucky with the weather or unusually good seeing? No! Like most things in life it all boils down to hard work and an almost obsessive dedication. As we have seen, the best amateurs observe on more nights and for longer each night. They also pay rigorous attention to telescope collimation, focusing, and a large image scale approaching 0.1 arc-seconds per pixel. Everything these amateurs do is pushed to the limit and every planet is treated differently: every planet has different color characteristics and different rotation rates. In terms of image processing, the top imagers leave no stone unturned. In passing, it is worth mentioning one constant source of confusion regarding the maximum exposure time that an unmodified webcam can offer. Virtually all commercial USB webcams like the Philips ToUcam Pro have a Windows menu structure that allows you to set a maximum exposure time of 1/25th second and a maximum frame rate of five frames per second. It might, therefore, be imagined that if you are in manual exposure mode and you force the webcam into taking 5 or 10 frames per second that the CCD is idle for most of the time between exposures. In fact, this is not the case. In manual exposure mode, with the exposure set to maximum (1/25th on the menu), a frame rate of 10 frames per second will give you exposures of 1/10th of a second and a frame rate of 5 frames per second will give you exposures of 1/5th of a second: the webcam will expose for the maximum time possible in manual mode, even if the 1/25th setting makes you think otherwise! I have carried out some tests on Philips ToUcam webcams and I came to the following conclusions regarding stated and actual exposures. On the maximum 1/25th exposure setting, frame rates of 5, 10, and 15 frames per second result in exposures of 1/5th, 1/10th, and 1/15th of a second. On the 1/33rd setting, frame rates of 5, 10, and 15 frames per second result in exposures of 1/10th, 1/20th, and 1/30th of a second. Finally, on the 1/50th setting, frame rates of 5, 10, and 15 frames per second result in exposures of 1/30th, 1/50th, and 1/50th of a second, respectively. In practice, the 1/25th and 10 frames per second settings deliver the best results, as 10 frames per second, delivers negligible image corruption and 1/10th of a second gives a good signal-to-noise, even with Saturn. Faster than 10 frames per second, USB 1.1 webcams are forced to severely compress the data into blocky chunks of 4×4 pixels, thus distorting information and, effectively, reducing the system resolution. But 1/10th of a second exposures are still good enough to freeze much of the atmospheric turbulence under good seeing conditions. Using frame rates higher than 10 frames per second *will* work, and stacking thousands of compressed and blocky images will tend to remove the blocky appearance nicely, but images still suffer. Under near-perfect seeing, a frame rate of 5 frames per second can be used to good effect, for a strong signal-to-noise ratio.

Dispersion

So far we have mentioned little about the color aspects of planetary image processing, but a vital understanding of color and the eye/brain perception of what appears on your monitor screen is essential.

From latitudes well away from the equator, the planets are never going to be directly overhead. This instantly causes a problem with respect to atmospheric dispersion, i.e., the splitting up of colors into a spectrum. We have all seen the way in which a prism splits light up into its constituent colors. Well, the Earth's atmosphere causes the same effect: the lower the object's altitude, the worse the dispersion. The effects are especially notable on the Moon, where bright crater edges will be fringed with red and blue. Unfortunately, atmospheric dispersion is significant enough to severely limit a telescope's resolution on any planet lower than 35 degrees altitude. A planet at 90 degrees altitude, that is, directly overhead, will have no color dispersion. At 60 degrees altitude (a typical best case scenario for observers in the U.K.), the visual spectrum from red to blue will be smeared across 0.35 arc-seconds, i.e., the theoretical resolution of a 30-cm telescope. Move down to 45 degrees altitude and the visual spectrum will be smeared over 0.6 arc-seconds, i.e., roughly the resolution of a 20-cm telescope. Things then get dramatically worse! At 30 degrees above the horizon dispersion will be 1 arc-second and at 18 degrees, 2 arc-seconds. Of course, at these low altitudes there will be other undesirable effects, too, as the light is coming through a lot of air, seeing will suffer and the image will look dimmer.

Fortunately, for the webcam imager, there is a partial solution to atmospheric dispersion. All digital images are constructed from red, green, and blue values, which can, at the user's discretion, be separated into their respective channels. For example, Registax has a feature called RGB shift (Figure 8.1) in which the user can

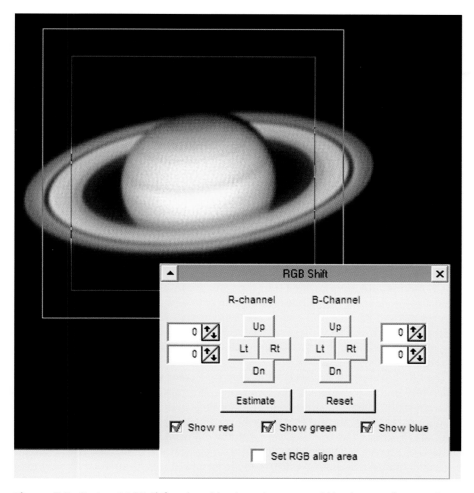

Figure 8.1. Registax' RGB Shift tool enables the red, green, and blue layers to be moved to compensate for atmospheric dispersion.

choose to move the red, green, and blue components of each image with respect to each other until no color fringes are seen at planetary limbs or around bright craters. Of course, this is not a perfect solution, but, aesthetically, a planet without blue and red fringes on opposite edges, looks much better. Needless to say, when a planet transits the local meridian (due south from the northern hemisphere and due north from the southern), it is at its highest point and this is the point of least dispersion. Another solution to dispersion is to use an optical arrangement by which prisms reverse the damage inflicted by the atmosphere. This might seem like a horrendous optical problem, but, in fact, AVA (Adirondack Video Astronomy) has recently marketed an affordable wedge prism corrector that can be set to correct atmospheric dispersion at a variety of altitudes. I remember seeing such a device in 1984, when I visited the legendary optician Horace Dall in his home, but now such devices are available commercially.

LRGB

An alternative to an optical wedge prism solution and to digital RGB channel splitting is to image the planets through narrow-band or separate red, green, and blue filters and combine the image into its final color solution. This might seem like a backward step when color webcams are cheap and easily available and, indeed, it is certainly more hassle. However, it has big advantages when a planet is at a low altitude. Color webcams have an array of filters covering the CCD pixels, from which they deduce the final color. As we have seen, in a typical example every 2×2 block of pixels might have a red-filtered pixel, a blue-filtered pixel, and two green-filtered pixels. Thus, the sensitivity of the webcam is significantly reduced by the filters. In a monochrome webcam, full-size glass filters have to be purchased to create the colors, reducing the sensitivity as before. However, an additional "luminance" signal can be taken with a monochrome camera, with no filters in the way, and this luminance signal will have much better signal-to-noise than the color images or the color webcam. The technique that is used to combine the excellent luminance signal with the color information is known as LRGB (luminance-red-green-blue) and it is a very powerful technique indeed. In fact, Registax has an LRGB mixing tool incorporated into the software that enables the color channels from a color webcam with a Bayer filter matrix to be split up and recombined (see Figures 8.2 and 8.3) but this not as powerful as true filtered LRGB with a monochrome webcam where we typically have 24-bit color (3×8 bit) information to play with. We will see shortly that even more powerful luminance techniques can be created with specific filters.

The LRGB feature in Registax, when used on a color webcam AVI, works slightly differently in that it constructs the luminance (brightness) signal of the final image from a mixture of the red, green, and blue channels in the existing image. The default ratio for the mix, i.e., the one that gives the most natural appearance from a ToUcam webcam, produces a luminance value with the following ratios: luminance = $0.299 \times$ red + $0.587 \times$ green + $0.114 \times$ blue. Not surprisingly, green is given the highest weighting as the human eye is most sensitive to green, and blue is given the least weighting as the blue end of the spectrum has already been boosted to noisy levels by the webcam's YUV image compression system. Altering the red, green, and blue sliders in Registax can make significant changes to the appearance of a planet imaged by a web-

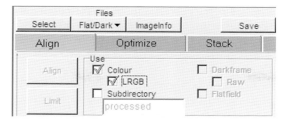

Figure 8.2. Checking the LRGB box in Registax enables the luminance (brightness) component of the final image to be constructed from the least noisy color channel.

cam. With Mars, increasing the red content significantly increases image contrast, whereas with Saturn, increasing the blue content (although noisy) enhances the subtle Saturnian belts. However, true LRGB imaging can only be achieved using filters on a monochrome camera. A cheap way of minimizing dispersion, for the complete beginner, is to employ a UV-IR block filter that at least restricts the spectrum to the visual band and knocks out the near infrared and ultraviolet regions. However, there are better solutions when monochrome cameras are used.

The Deep Red Advantage

At first it might not be clear how the LRGB technique can be applied to curing atmospheric dispersion. Surely, in the luminance image, the full dispersive smearing is still captured, but is it simply a monochrome smearing? Well, this is true, but

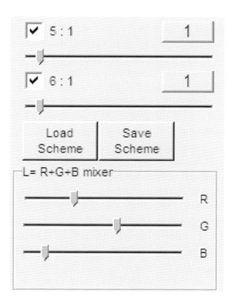

Figure 8.3. Once LRGB processing is checked, the LRGB sliders in Registax allow the luminance signal to be constructed from any mixture of the color channel signals.

amateurs have developed some powerful methods for coping with a planet that is low down. For a start, the luminance signal need not be derived from an unfiltered monochrome image, or even a UV-IR rejected image. A narrow-band filter that captures the maximum contrast on a specific planet can be used to dramatic effect. If this narrow-band image is taken in the deep red, the results can be very dramatic indeed. To anyone who knows about optics this might seem counterintuitive. After all, blue light has a smaller wavelength than red light and a smaller wavelength means higher theoretical resolution, right? Wrong, in this case, basic theory and practice do not line up. Atmospheric seeing is far, far better in the deep red even if the wavelength is longer! A near infrared filter is the closest thing you can get to a seeing filter. If you can take a luminance frame in the near infrared, the resolution will be superb, especially on high-contrast objects like the Moon and Mars. Remarkably, Jupiter also responds well. CCDs are especially sensitive in the red end of the spectrum and less light is scattered in the red, too. If you do not care about color information, then things are really looking up. Near-infrared monochrome shots of the Moon, Mars, and Jupiter can all look breathtaking, even when the object is at a fairly low altitude.

If you do want a color image, then you can indulge in some pretty clever techniques to exploit the extra resolution gained in the deep red. This is especially powerful on Mars, where the planet is mainly red or pink in appearance anyway and using red as a luminance image will not introduce any grossly inaccurate color effects. Damian Peach has produced some stunning color images of Jupiter at only 30 degrees altitude by imaging the planet with a monochrome AtiK webcam through deep red (for the sharpest frames) and blue filters. To synthesize green he just averages the deep red with the blue signals. The resulting images look identical to RGB images, except green was never used! In the case of Jupiter, this saves valuable time as Jupiter rotates quickly. As far as I am aware, this "synthesis of green by averaging red and blue" technique was first pioneered by Antonio Cidadao. I still find it amazing that a full-color image can be obtained from just red and blue! However, there are not many green planetary features, so synthesizing green is probably not a critical test. If, to gain extra sharpness, the luminance signal is biased more toward the red, some color imperfection can be expected, but this is a small price to pay for a sharp image. In practice, Mars' color does not suffer much from a red luminance, Jupiter does, and the luminance image for Saturn is often chosen as green or green + red, but different imagers have different techniques.

At this point the reader might become just a bit confused at exactly what color is all about. Surely, there are only red, green, and blue subpixels on a computer screen? Where does the L value come in? Is there an extra L subpixel? No, there is no L subpixel. PC screens are made from an array of red, green, and blue subpixels, but the colors you actually see are all to do with the relative *ratios* of red, green, and blue subpixels. However, the luminance (brightness) you see is concerned with how brightly those subpixels are glowing. You can have a yellow object (red and green subpixels equal in brightness, blue subpixel turned off) but if the luminance value is low for that pixel, then all the subpixels will be dimmed by the same amount. That luminance (brightness) value can be derived from any narrow-band filter you like. However, despite the fact that luminance determines the perceived brightness/contrast between different pixels and not the color, it can skew the perceived color dramatically. After all, if you imaged a red planet like Mars through a blue filter and used blue for the luminance information the planet would appear virtually black. The RGB information might say it was still a red planet but an almost black red planet does not look like the bright red/orange Mars we are used to! Mars, as I have said above, is the best planet to use a deep red filter with because not only is the planet red anyway, seeing is always better in the red. An LRGB image of Mars with the L component derived from a deep red filtered image can give a pretty true color rendering.

Blue is a real problem area in planetary imaging and not only because of the poorer seeing and greater atmospheric scattering in the blue end. The techniques used to compress webcam information, on the journey from webcam to PC, also favor the red colours, as does the CCD sensitivity itself. Even when imaging deep sky objects, filtered LRGB exposures often require twice the blue exposure as the red images. In fact, LRGB was originally invented for deep sky and not planetary work. Because filtered deep sky exposures are so noisy with respect to unfiltered ones (because so much light is lost with already faint objects), it was realized that to get a deep high-resolution image of a galaxy or nebula it was always best to take an unfiltered monochrome image: the signal-to-noise was just so much better. However, it was then realized that if you literally color the resulting low-noise monochrome image with the color ratios from the noisy color image you get the best of both worlds: a clean, deep image with added color. At the pixel level, the ratio of brightness between the red, green, and blue subpixels colors the scene, and the depth of the monochrome image provides the low noise. In fact, the eye-brain perception of color helps here, too, because the eye is very sensitive to luminance resolution and far less sensitive to color resolution. So insensitive, in fact, that filtered images for the deep sky color information can be taken at half the resolution, i.e., pixel info can be binned 2×2 so that 4 pixels contribute to the light measured. Indeed, for deep sky work the color information can even be provided from a smaller telescope.

The LRGB technique is not quite as transforming in this respect for lunar and planetary work because, lets face it, the Moon and planets are very bright objects. But where the technique is at its most powerful here is in allowing deep red, narrowband, and nonblue filters to be used as the luminance signal, at colors where seeing is better and a planet has more detail. However, the LRGB technique does have a big disadvantage for planetary work. Namely, planets rotate, and, in practice, filters need changing in a short period of time. This can also necessitate some refocusing work, and everything has to be completed in a few minutes. If you can get by with just two filters, i.e., deep red and blue, in a reliable filter wheel, things need not be too fraught, but beginners will certainly prefer a one-shot color webcam approach on objects at a decent altitude, where serious dispersion and seeing problems are not an issue. One technique I have used to good advantage when imaging Mars is the two-webcam approach. Mars is a small planet with a much slower rotation period than Jupiter and Saturn so there is plenty of time to take both a deep red AVI video with a monochrome, filtered webcam and a color video with a ToUcam Pro webcam. However, using a sensitive monochrome webcam and proper color filters is best. Commercial low-profile filter holders are widely available (Figure 8.4) or you can make your own, as I did (Figure 8.5). When buying a filter set I would advise purchasing green and blue filters with an infrared rejection coating, but a red filter without an infrared rejection coating, as well as

Figure 8.4. A commercial, lightweight, low-profile, manual filter wheel attached to an ATiK webcam. A low-profile system is of particular advantage to the Newtonian user where the light cone is fixed. Image: Jamie Cooper.

Figure 8.5. An ultra-low-weight home-made filter holder built by the author and his father for the author's 250-mm f/6.3 Newtonian. The filters are mounted in 13-mm-thick slabs of Perspex and slide down channels in a groove cut in the side of a 50-mm drawtube.
Image: Martin Mobberley.

an infrared (I band) filter (from 700 to 900 or 1000 nanometers). The infrared (I band is 700–900 nanometers) filter is incredibly useful for low-altitude imaging of the Moon, Mars, and Jupiter and the non-IR blocked red filter is great for providing a strong red signal on high-altitude objects. CCDs are very sensitive in the infrared and it is a shame if the infrared component is not used in filtered work. Conversely, with a color webcam, a UV-IR blocker is recommended for low-altitude work to restrict the effects of dispersion. The effects of atmospheric dispersion on a low-altitude Mars are clearly shown in the ToUcam image in Figure 8.6. A huge improvement is gained by using an infrared luminance filter and realigning the color layers as seen in Figure 8.7. Saturn seen through red (non IR blocked), green (IR blocked), and Blue (IR blocked) filters is nicely shown in Figure 8.8. Figure 8.9 shows the Maxim DL software tool for creating LRGB images.

Gamma

Many terms used in planetary imaging are already in frequent use by nonastronomers. The digital camera era has plunged many otherwise sane and normal people into the world of image-processing jargon. However, the term *gamma* is rarely understood and it is so crucial to planetary imaging that an explanation is required. At first glance it would appear that increasing the gamma of an image makes it brighter and reducing the gamma of an image makes it dimmer. However, it is a lot more subtle than that. Play around with the gamma function

Figure 8.6. The color dispersion on a low-altitude (20 degrees) Mars is obvious in this image by the author, taken in August 2003. A red/yellow fringe to the south (top) and a blue/violet fringe to the north (bottom) reduce the achievable resolution. Image: Martin Mobberley.

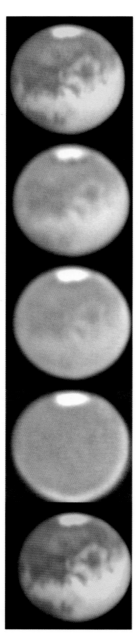

Figure 8.7. Mars (from top) in infra-red, red, green, and blue light (these are sharpened images.) The final image is an LRGB composite employing the clean luminance data from the infra-red image and the colors from the low-res, noisy, red, green, and blue images. Image: Martin Mobberley.

Figure 8.8. Saturn, in near-perfect seeing, imaged through red, green, and blue filters with a Celestron 9.25 SCT at f/40 and an ATiK 1HS webcam. The blue image is the noisiest. Image: Damian Peach.

on any image-processing package like Photoshop or Paint Shop Pro and you will notice that the brightest and darkest parts of the image stay the same, whereas the mid-range brightness varies considerably. The gamma function makes use of an interesting property of numbers between zero and one, raised to a certain power. In PC image processing, one way of representing brightness is by regarding jet black as zero, brilliant white as 1.0, and the mid-range brightness as 0.5.

As a slight digression, inside the computer's processor the brightness of a pixel might be represented by 8 bits (0–255) or 16 bits (0–65,535). As an example of an 8-bit number, imagine the binary number 00010000. The right-most digits in 00010000 that represent 1, 2, 4, and 8 are all zero as are the left-most digits representing 32, 64, and 128. However, the digit representing 16 is one, so 16 in decimal = 00010000 in binary. The full range of pixel brightnesses (from 00000000 to 11111111 in binary) are equal to 0 to 255 in decimal. But, as far as gamma is concerned, 0 = 0, but 255 = 1. Everthing is scaled down such that black is 0 and white is 1.

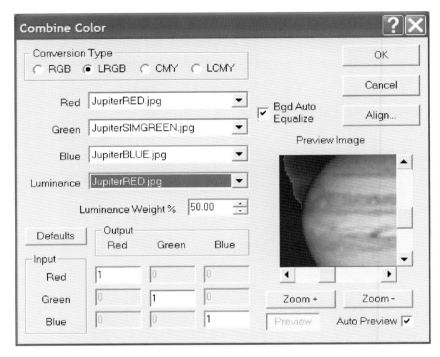

Figure 8.9. The LRGB tool in Maxim DL. In the example, an LRGB image of a low-altitude Jupiter is being created using infrared as red, blue as blue, a simulated green (red plus blue averaged), and an infrared luminance with a 50% weighting. In other words, the end result is a sharp tri-color Jupiter using just two filters!

Anyway, back to the plot. The clever bit is that numbers between 0 and 1, whatever power they are raised to, stay between 0 and 1. This is the essence of the gamma function. In addition, 0 raised to any power stays at 0 and 1 raised to any power stays at 1.

The gamma function is generally represented by the following formula:

final pixel brightness = original pixel brightness raised to the power 1/gamma, where the original pixel Brightness is between 0 and 1. Or, FPB = OPB$^{1/\gamma}$.

If gamma (γ) is set to 1, then FPB = OPB, i.e., there is no change in brightness. However, if gamma is less than 1, the mid-range brightness will be dimmer than before; if gamma is more than 1, the mid-range brightness will be brighter than before.

Using real numbers, with the original mid-range brightness of 0.5:

With a gamma of 0.7, FPB = 0.5$^{1/0.7}$ = 0.37, i.e., brightness has dropped from 50% to 37%. With a Gamma of 1.3, FPB = 0.5$^{1/1.3}$ = 0.59, i.e., brightness has increased from 50% to 59%. But in each case, the minimum, 0%, and maximum, 100%, brightness will not change.

At this point, the reader may wonder why I am laboring this mathematical point so much, especially as the brightness variations in my example appear so trivial. However, when you see the effects of a gamma variation on a planetary image your opinion may well change! Spherical planetary objects with subtle features can show significant improvement when the gamma is reduced. The effect is especially notice-able with Jupiter, where it is advisable to set the webcam's own gamma very low to start with. While it is advantageous to retain the darkest and brightest (e.g., Jupiter's equatorial zone) features on a planet at their original values, reducing the brightness of the mid-tones reveals substantial details in Jupiter's brighter regions; details that, with a high gamma setting, would seem over-bright and washed out. The advantage of a gamma reduction (combined with unsharp mask/wavelet processing) means that the maximum fine-detail contrast can be achieved without the planetary limb disappearing or the planetary equator whiting out. Registax has a built-in gamma function tool for adjusting the gamma of the final image (Figure 8.10).

Altering the gamma on Saturn can enable fine gradations in the polar regions to be captured, while not saturating the equatorial region and maintaining a good brightness to the rings. In the era of photography, two of the biggest planetary has-sles were preventing Jupiter's limb regions from darkening and disappearing and stopping Saturn's rings from doing the same. With careful tweaking of brightness, contrast, and gamma in Registax, both features can be saved from disappearing into the blackness, and a nice, punchy, high-contrast image can be preserved, too.

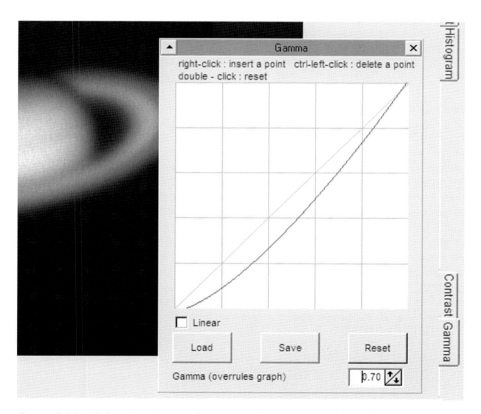

Figure 8.10. Clicking the Gamma tab in Registax brings up a useful graph showing how much the mid-range brightness of the final image is being altered.

Preserving the Jovian limb is essential when trying to measure the positions of features on the globe.

Raw Mode

In recent years a technique known as "raw mode" has been popular among webcam planetary imagers who like to push their equipment to the limit. To understand raw mode we need to investigate in more detail how webcams transfer so much data so quickly. Most webcams use the USB 1.1 system to transfer their video streams from the webcam to the PC. USB 1.1 has two data rates, namely, 1.5 megabits/second and 12 megabits/second. The slower rate is reserved for simple hardware like keyboards and mice whereas webcams use the 12 megabit/second rate. Webcams typically have CCD arrays of 640 × 480 pixels, i.e., 307,200 pixels and they use a color-encoding technique called YUV to compress the color data from the filtered pixels. This compression technique works on blocks of 4 × 4 pixels and the letters stand for the following: Y = luminance, U = red–luminance; V = blue–luminance. The YUV technique is very efficient at transmitting acceptable color images of everyday scenes from a webcam to a PC. A quick calculation shows that at 8 bits of data per pixel all of the monochrome luminance information from a 640 × 480 grid could be transmitted at 5 frames per second without any data compression. In fact, with YUV, color images look remarkably uncorrupted even at 10 frames per second. However, even at 5 frames per second, the YUV compression does take subtle liberties with an image and the data is not as clean as it would be with a monochrome camera. In addition, control of the webcam's gamma response and artificial sharpening algorithms are used to give a more acceptable picture at high data rates for daylight scenes with the webcam lens. However, the planetary imager is not using the webcam in normal mode; he or she wants the most accurate transmission of data from webcam to PC. For many planets, especially Jupiter, the webcam gamma setting is best turned down to a low value and a frame rate of 5 frames per second is best when seeing is excellent. The planetary imager is interested in low noise frames, too. Because of all these considerations, various software "gurus" worked out how to run a routine that would reset the standard domestic webcam to its default factory settings in manual mode with a slow frame rate, low gamma, and sharpening algorithms turned off. The end result is a much smoother raw planetary image but with some additional processing required. In most variants of the raw mode, the webcam images download in monochrome. They are much cleaner images than the color ones, but are, essentially, the raw output with no attempt to reconstitute the color information. Thus, they will be corrupted with the filter grid (Bayer Matrix) pattern. However, your PC can reconstitute the color for each frame (and the thousands of frames in each webcam video) with the right software. Various amateur software enthusiasts offer freeware to download and make the raw mode possible on your PC, but with a disclaimer that the risk of damaging your webcam is yours! (See http://www.astrosurf.com/astrobond/ebrawe.htm) Is raw mode worth it? This is a tricky question. Certainly, for the perfectionist, who has squeezed every last drop out of his or her camera, the raw mode will make a subtle improvement to the highest quality images. But, for the beginner, the extra hassle involved in reconstituting the color image will not make any noticeable difference until all the other issues like collimation, focusing, and prolific observing have been addressed. Many monochrome webcams are, by default, working in raw mode, so if you use these, switching to raw mode is not necessary.

CHAPTER NINE

Advanced Stacking of Rippling AVI Frames

We have already examined the remarkable Registax software developed by Cor Berrevoets, a software package that has made the stacking of thousands of AVI frames a routine and reliable operation. However, few amateurs who use Registax seem to have the spare time or the inclination to examine exactly how it works and to assess which options are the best in different situations. However, a full understanding of Registax' strong and weak points can be of great use. As a Registax user myself, I hope that an explanation of the software from a user's (rather than the designer's) viewpoint will be of use. I would certainly have valued a Registax "Idiot's Guide" when I started using it!

One thing that needs to be appreciated from the outset is that even on the best nights of atmospheric stability a planet never just sits there, without a single quiver. In the webcam era it is now possible to video a planet for many minutes, at 0.1 second intervals, and to see just exactly what happens from split-second to split-second. This can be highly revealing. Remember, the light is traveling through 30 kilometers of the Earth's seething atmosphere. It would be incredible if the image of a planet were to remain steady and undistorted for several minutes at a time: indeed, it never happens! What does this imply for a frame-stacking program? It is important to remind ourselves why we stack frames at all. Stacking reduces the pixel-to-pixel noise by a factor roughly equal to the square-root of the number of frames stacked. In addition, on images of reasonable quality (i.e., not highly distorted) it produces an average image, not one where one side of a planet, or its rings, is deformed. Of course, a few, rare, individual frames may well be nearer to geometric perfection than the final result, but they will be far noisier. Stacking also increases the dynamic range of an image such that subtle shades and contrast differences in one noisy 8-bit frame are transformed into a smooth final result where the most subtle luminance variations are revealed. Out of several thousand frames there will always be a few single, noisy frames, where the seeing was close to perfection for that single 0.1 second period. In those frames the planet's shape may well be within an arc-second of the true shape all around the limb or rings if seeing conditions are good. Frames such as these are the master reference images that Registax needs to use to align and stack other good images onto and this is the key to how Registax delivers the goods. Needless to say, the master reference frame must be chosen with care. (In fact, when you become more experienced, a master reference image, produced using Registax' Optimizing page "Create" function can be assembled from a short sequence of 50 frames, rather than a single, noisy, frame.)

Once you have a master reference frame you will then want to align and stack as many good frames as possible, with respect to that master frame, to reduce the noise. The fact that the planet wanders around a bit at the whim of the telescope drive is not a disadvantage here (as long as it does not wander too much). A major factor enabling noise reduction is that there is often a subtle fixed pattern to noise for each CCD chip. With long exposure CCD images, a dark frame or bias frame is often taken to eliminate this fixed pattern noise. But, with planetary webcam frames, the pattern is no longer fixed if the images are stacked in a random fashion (as random as the planet's movements). Even stacking hundreds of crude 8-bit dynamic range webcam frames increases the bit depth of the image noticeably.

Of course, if you stack thousands of images the final result will look incredibly smooth. However, you really do not want to stack those images that are blurred or badly distorted. So here you have a dilemma of a sort. Where do you draw the line? From a typical webcam run on, say, Saturn, you might collect 3,000 frames in a 300-second period. Twenty of those frames might be very sharp and undistorted. Several hundred might be close to being sharp and just a bit distorted. The rest will be blurred and distorted in varying degrees. How do you (or Registax) decide whether to stack many (for a smoother result) or few (for a sharper, but noisier result). There is no simple answer to this, except to say that experience plays a big part, every planet is different, and Registax has numerous settings to help you get the pass/fail criteria correct. Excellent webcam images of the Moon can be assembled from a few dozen frames. This is because the Moon is a bright, high-contrast object and intensive image processing is not needed to bring out the smallest craterlets and rilles. Also, distortion is far more obvious across a large lunar landscape: double images of high-contrast features stand out. Distortion on a lunar image will become worse as you move further outside the Registax alignment box, unless seeing was near-perfect. With the globe of Saturn any features are usually so subtle that they will only reveal themselves when a thousand or more webcam frames (from a 10 frames per second and f/30 – f/40 system) are stacked. So, for the relatively faint ringed planet, stacking loads of frames is infinitely more important than with the Moon.

The first Registax page, or window, allows you to view every frame of your webcam AVI video (or individual bitmap, jpeg, tiff, FITS, or png images). To load your webcam AVI you simply click on "Select" and then load the AVI file. Right clicking on the best image on the center of the planet with a suitable alignment box size will set that good image as the reference. The features inside the alignment box will be used as the reference to mathematically align the other frames with respect to, for example, an entire planetary globe or a lunar crater. To choose a really good reference image you really need to open the Registax frame window and chug through a lot of frames manually until you spot a really good single frame. You can then stack hundreds or thousands of frames with respect to that one good frame (or a composite of good frames). But how, precisely, do you make the decision as to which frames to use in the stack and which to reject? Well, if you have an excellent image and really want to be meticulous (or if you are just an unbelievably sad loser with nothing else to do) you can simply plough through each AVI frame visually and tick (select) all the ones you want. Obviously, if you have thousands of frames this will take some hours to do! But, you can also get Registax to help you. In other words, give the software a few clues as to what your thresholds are and get it to check each frame for quality while you go off and do something less sad and tedious!

Let us examine this process in more detail, because it is crucial to our understanding of Registax if we eventually hope to fully master the software. When you select a reference frame you need to specify an alignment box size. Registax gives a choice of 32, 64, 128, 256, or 512 pixels for the box size. Most planetary imagers simply choose a box to surround the globe of the planet. Once the alignment process is running, if the telescope's drive causes the planet to wander completely outside the chosen box position the software will get confused and may prompt you to manually register that frame. At the start, once you have clicked your mouse on the planet or the alignment feature in the best image, Registax performs some calculations, presents you with some data, and leads you to the alignment page. A warning here: if the next few paragraphs seem totally incomprehensible that is what I would expect if you are a novice user of Registax! The finer points of what Registax is doing will largely be a mystery until you have played with the software over weeks and months. There is quite a learning curve for the beginner using this software. However, the good news is that you do not need to understand Registax in detail to get a good result. The default aligning and stacking system works well; you only need to understand the software in detail if you become a perfectionist. Anyway, to continue: at the alignment page (Figure 9.1) the data you are presented with at this stage is a colorful picture called an FFT Spectrum and a graph labeled "Initial Optimizing Run" (or "Registration Properties" in Registax versions 1 and 2). The colorful picture should show a small red circle at its center if the software's initial estimate of the shift between reference and other frames is good. The FFT alignment value in pixels can be altered and "Recalc FFT" pressed if a nice small red circle is not initially seen (although I rarely have to use the Recalc option). The Registration Properties graph on the alignment page shows a red line, which is an indication of the "Power Spectrum"; in other words the relative amount of large and small features in an image. The Quality tab (Figure 9.2) will, at this point, bring up a Quality Settings box that, in conjunction with the Quality Estimate Method, helps define the image quality assessment in conjunction with the Power Spectrum. (I can see your eyes glazing over already . . . hang on in there!) Early versions of Registax have just one method of quality assessment, referred to as the "Classic' Method" in Version 3 and later. The quality assessment methods are called "Classic," "Human Visual," "Compression," "Local Contrast," and "Gradient." In "Classic" mode, the two green lines on this graph define a quality band, which the user can alter the position and width of. In "Human Visual" and "Compression" modes the user can also alter the quality settings. Cor Berrevoets added the "Gradient" quality assessment after version 3 trials by Anthony Wesley

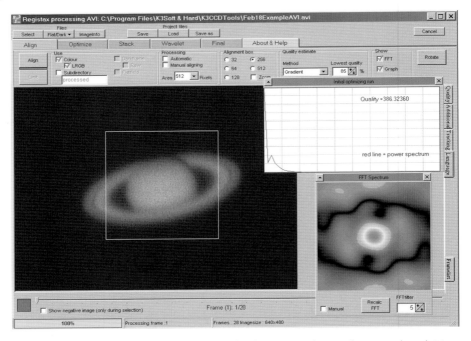

Figure 9.1. The alignment page in Registax after the master reference frame is selected. Note the FFT and initial optimizing run windows have appeared.

and Damian Peach. Damian, in particular, did not like the way Registax ranked the image quality more by shape than by sharpness, especially on frames taken under near-perfect seeing conditions. For really good, clean, sharp images, the "Gradient" quality assessment works best.

Confused? I would be surprised if you were not! As I said earlier, letting all this sink in will take quite a few trials (dozens!!) at using the software. In a nutshell, the Power Spectrum line, FFT Spectrum, and Quality Settings choices are all there to try to determine how to quality grade your images and ultimately accept or reject them. If we choose a Quality Setting of "Classic" as used in all the early versions of Registax, we can shift the vertical green lines left and right to choose our quality band along the red Power Spectrum curve. If we set the left-hand green line at the intersection point where the red Power Spectrum first starts to flatten out after its downward plunge, and the right hand green line at the intersection point just before the red line hits the bottom, this will do nicely. By doing this we are choosing to assess the quality of an image based on how much medium and fine scale detail is present, but not on fine scale noise (the bottom of the graph). In Registax version 3 and later versions the "Human Visual," "Compression." "Local Contrast," and "Gradient" quality assessment methods were added, although the first two do not work well for me. They seem to grade an image much more on its physical shape with respect to the reference master than on its sharpness. Using these methods might lead to less artifacts being produced in Saturn's rings or on the Moon, perhaps, but they will generally lead to less detail being seen in the final

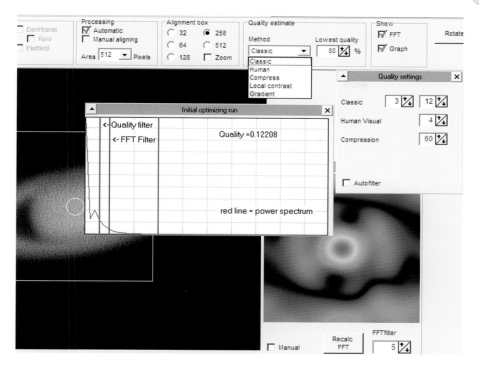

Figure 9.2. Quality estimation in Registax. Registax has five ways of judging and ranking the quality of individual frames, depending whether geometric distortion or sharpness are the most important factors. These are called Classic, Human, Compress, Local Contrast, and Gradient. Classic, Human Visual, and Compression have extra settings that can be adjusted. A graph enables you to see the adjustment to the position of the green quality bars in classic mode, set to 3 and 12 in this example.

image. At least, that has been my experience; although my tests are certainly not exhaustive.

Although we have been talking about image quality assessment we have not yet talked about setting the accept/reject quality threshold. This value is seen in Registax' "Quality Estimate" window. Altering the percentage value in this window tells Registax what your accept/reject threshold is, although, admittedly, a percentage on its own does not mean much to the novice. Some experience is needed before this value can be set with confidence. The default value is 80%, but if you are sad enough to have manually de-selected all the poor frames already (yep, I've been that sad on quite a few occasions), 50% works well.

Once you have set all the values you want on the alignment page it is time to actually click the "Align" button for the initial alignment of all the frames or images. Once this initial alignment has run its course, pressing the "Limit" button (which limits you to the frames above the determined quality threshold) takes you to the "Optimize" page. During alignment, Registax places the files in order of their quality and the slider tool at the base of the page stops at the quality threshold. Images to the left on the slider are high quality and those to the right are low quality. This is a new feature from version 3 onwards. You can alter this slider to the image threshold you prefer, remembering that images to the right will be rejected in the final stacking process. Once you press "Limit," the threshold that you (or Registax) set is determined and you can then proceed to the Optimizing or Stacking pages.

Optimizing (Figure 9.3) is, essentially, a precise and time-consuming optimizing of the image alignment. The "Search Area" and "Optimize Until" boxes are the ones that will most concern the beginner here. Registax does have help files for all

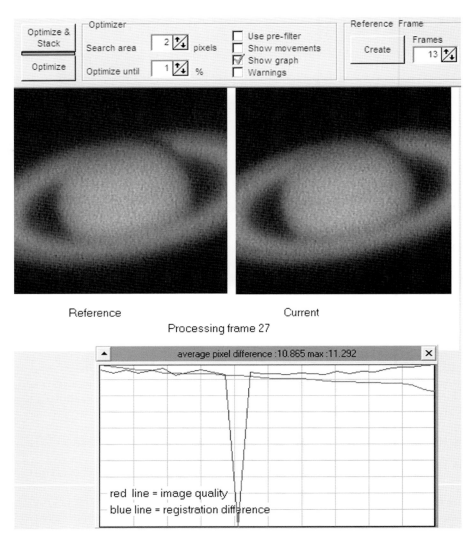

Figure 9.3. While optimizing the alignment process (an optional step) the master image and frame being processed are displayed, along with a graph showing the alignment errors and average pixel differences compared to the master frame.

of its features. However, my intention here is just to get the new user up to speed and this section is written from my perspective—an end-user with no knowledge of the software design. I think with all software packages the designers are rarely able to see things from the perspective of the user, i.e., from a perspective of total ignorance. In fact, I know this is the case because writing software was my profession at one time. My intention is to allow the beginner to surmount the initial learning hurdle to the point where Registax is not so intimidating that it stops the beginner dead in his or her tracks! The "Search Area" and "Optimize Until" boxes are there to allow the software to search an area of a number of pixels and to continue the iterative optimize process until the percentage change in shifts for all the images is less than a certain value. Choosing to search a larger number of pixels for a best match will take more time but may mean less iterations taking place. In fact, you do not have to optimize at all. With noisy images, and, especially, with a specific noise pattern (e.g., from having a telescope controller hand paddle close to the PC/webcam) Registax can lock onto the noise pattern when optimizing. This seems to happen often with noisy and low-quality images. In this case, optimizing leads to virtually no improvement in the planetary image but a large increase in noise. This increase is not obvious at first, until the wavelet-processing/enhancing stage, when noise can really emerge and become unsightly. I would strongly advise keeping all sources of electrical noise away from the PC webcam when imaging. Unfortunately, the PC is a major source of noise itself and some PCs appear to be better shielded than others. If you always get noise problems it may be worth trying another PC when imaging. At long focal lengths I have not experienced any significant problems when leaving the alignment stage out, if noise problems emerge due to optimizing. However, if noise problems do not emerge I keep Optimize turned on. With a slow PC, i.e., under 1 GHz processor speed, turning Optimize off can really speed things up, meaning you wait half an hour, not several hours for the aligning, optimizing, and stacking process. In the worst case, with a slow PC sorting thousands of poor frames, the optimizing process can take half a day to complete. In these instances I generally leave the PC running overnight, while I am sound asleep in bed!

After the optional optimizing process (or, as part of Optimize and Stack) we come to the stacking process, during which stage all our best images are simply stacked and averaged resulting in a supersmooth, if less-than-sharp, image. The stacked image tends to look very bright because the stacking process makes sure that the brightest point of the stacked image is just below 100% saturation. At this stage the only task remaining is to apply the powerful wavelet processing routines on the Registax Wavelet page.

Final Processing

The Wavelet page is, undoubtedly, the Registax page where really excellent images are created. This is where you can endlessly "tweak, tweak, tweak, tweak, tweak" the image until you are diagnosed as having an obsessive compulsive disorder and the men in white come to take you away. What in God's name is a wavelet anyway? Well, the term *wavelet* originates from the world of DSP, or digital signal processing. In electronics, as well as in planetary imaging, the challenge is often to extract signal from the noise, an especially tricky task when the noise frequency and signal frequency are virtually the same. This is where long focal lengths help, provided you have sufficient light grasp. If your finest planetary details span numerous pixels they will have a lower "frequency" than the finer pixel-to-pixel noise. Thus, noise can be suppressed while detail is enhanced. We are in the world of mathematical transforms here where signals can be reduced to digital numbers and mathematical series (like sine and cosine). Fortunately, we do not need to get involved with the math. Thank goodness for that! Registax' genius inventor, Cor Berrevoets, has done that for us. The wavelet layer sliders on the Wavelet page correspond to increasingly coarse detail as we move from layer 1 to layer 6. The default "Initial" and "Step Increment" values of 1 and 0 can be used as well as the default wavelet filter and wavelet scheme (linear) settings. The first time you use Registax you can leave all the clever bits at their default values; you don't need to understand what they do to use the software and get a good result. Once you get to the Wavelet page, if you are taking planetary images at a nice, large image scale of around 0.1 or 0.2 arc-seconds per pixel you will find that using the sliders for layers 1 and 2 just increases the image noise and nothing else. So you can leave these untouched. Indeed, it is at this stage that you will discover whether the optimizing routine has enhanced any electrical noise pattern in the image. You may want to scrap the wavelet processing at this stage and stack the images all over again, with "Optimize" turned off. Sometimes this can reduce a noise pattern, but if noise is not a big problem leave "Optimize" turned on.

How does wavelet processing compare with traditional unsharp mask processing? Well, the effects are remarkably similar although the Registax layer sliders system is a much slicker, more powerful system. With the original photographic unsharp mask technique, a blurred version of the whole image was used as a filter to project the original negative through. The filter, being blurred, suppressed low-frequency bright information (such as Jupiter's white zones) revealing higher frequency, more subtle data, which was previously hidden. In the digital unsharp mask a radius (in pixels) is specified which tells you how blurred the mask will be. A strength is usually specified which is equivalent to the density of the blurred mask from the photographic era. The Registax layers are roughly equivalent to different unsharp mask radii, so all six layer sliders are like applying six unsharp masks of different radii to one image simultaneously—powerful stuff! Essentially, you endlessly tweak the layer sliders until you get the most aesthetic image. The trick is to know when to stop, as too much sharpening leaves an image looking noisy and unnatural.

There are other options on the Registax Wavelet page that are almost as important as the wavelet layer sliders themselves and can make a crucial difference to the final appearance of a planetary image. Apart from the simple brightness and contrast controls, the most critical options are the RGB shift and the gamma function. As we have already seen, in North America and northern Europe the Moon and planets are rarely at a decent altitude. When these objects are at their highest northerly declinations, i.e., above +20 Dec, they can achieve altitudes of 60 or 70 degrees from latitudes of 50 and 40 north, respectively. However, Mars, Jupiter, and Saturn spend half their time below 0 Dec and when below 40 degrees altitude atmospheric dispersion splits up the colors from these planets noticeably. Registax' RGB shift tool is a partial solution to this problem and it works very well, effectively realigning the red, green, and blue channels of the color image. Obviously this is not a complete solution: dispersion occurs within these color bands, too, and seeing is always worse at lower altitudes. Also, filtering the colors before they hit a monochrome webcam chip works far better. However, images look remarkably improved after applying the Registax RGB shift. Northern planetary limbs are no longer surrounded by a blue rim and southern planetary limbs are no longer surrounded by a red rim (and vice versa in Australia, South Africa, and New Zealand). Registax' RGB shift tool is pretty self-explanatory, with buttons giving you the option of shifting the red or blue channels, up, down, left or right a pixel at a time, until the planet appears not to be fringed with red or blue.

The final Registax page (Figure 9.4) contains a number of image tweaking features that, while useful, can be found on virtually any image processing package, such as JASC's Paint Shop Pro, or Adobe's Photoshop. These features allow altering the precise color hue and degree of color saturation (reducing color saturation can substantially reduce the noise, especially in a one-shot color image), resizing the image (useful for when Saturn's rings appear blocky), and rotating the image such that north (or south) is at the top. Most modern image processing packages also feature an auto-color balance feature, which attempts to give a scene a red/green/blue color balance that is correct for objects illuminated by sunlight. As everything in the solar system is illuminated by sunlight, auto-color balance can work well on the planets, too.

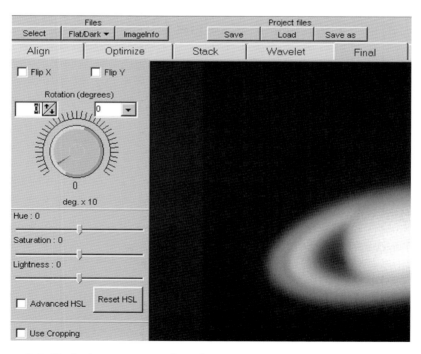

Figure 9.4. The final Registax page allows the image to be rotated and the Color hue, saturation and lightness to be adjusted.

CHAPTER TEN

Imaging the Moon

Without a doubt, the lunar craters and mountains are the objects on which all budding, high-resolution planetary imagers cut their teeth. The Moon is available all year round (though not always at a decent altitude), it is a huge target to aim at, and the contrast between light and dark regions on the lunar surface far exceeds the contrast available on any other planetary body in the night sky. It is also a rather brighter target than the gas giants Jupiter and Saturn, unless the most shadow-filled regions near the terminator are being considered. If we allocate a surface brightness value of 1 to Saturn, the relative brightness of the planets come out approximately as follows: Mercury = 80, Venus = 300, Mars = 15, Jupiter = 3, Saturn = 1, Uranus = 1/4, Neptune = 1/15, and Moon = 3 to 15 (half to full Moon). As we can see, despite its dark albedo (the Moon only reflects 7% of light falling on it) the lunar surface is a nice bright target for our webcams and it contains loads of fascinating objects to image. It is a real test bed on which to perfect our imaging skills.

North, South, East, and West

Before we start looking in detail at the Moon, I would just like to clarify the issues of up, down, east, and west! Standard astronomical telescopes, such as refractors and reflectors, always show the Moon upside down. Schmidt-Cassegrains are even more complicated as they often come supplied with a 90 degree prism that makes the eyepiece more comfortably placed, but creates a mirror image. However, historically, the Moon has always been sketched with south at the top, as seen through a reflector or refractor from the northern hemisphere. In this chapter, as in all the planetary chapters, south is at the top, i.e., the Moon is upside down compared to how it looks to a naked-eye, northern hemisphere observer. East and west can be confusing, too. Prior to the space probe era, features on the Langrenus/Petavius/Mare Crisium side were on the west and features on the Aristarchus/Gassendi side were on the east, as was the extreme eastern limb feature called the Mare Orientale, or "Eastern Sea." However, the International Astronomical Union (IAU) reversed this ruling in the space probe era and so Langrenus, Petavius, and the Mare Crisium are now in the east and Aristarchus and Gassendi are in the west. Mare Orientale is now on the western limb: yes, you've got it . . . the "Eastern Sea" is on the western limb of the Moon!

The Motions of the Moon

The Moon is unique among planetary bodies in that even from high-temperate latitudes, such as the northern USA or northern Europe, it can appear high in the sky every year. With planets that are further away from the Sun than the Earth, high nighttime altitudes are only achieved when the planet is reaching opposition in the winter sky, because then the Earth's polar axis is tilted toward the planet. Frustratingly, the best seeing conditions tend to occur in the summer, when the polar jet stream is far away and high-pressure weather systems are clear, not cloudy. With Jupiter taking 12 years and Saturn taking 29 years to orbit the Sun, waiting for a planet to be high in the sky again can be a frustrating business, unless you live near the equator, or are prepared to travel abroad. The Moon orbits the Earth every month and so it is always at a high northerly declination once a month and a high southerly declination two weeks later. This declination shift is largely due to the Earth's axial tilt of 23.5 degrees, but there is an extra 5 degree tilt (the angle the Moon's orbit makes with the ecliptic) superimposed on this 23.5 degrees. Thus, in the most extreme cases the Moon can actually range between +28.5 and − 28.5 declination during a month or, nine years later, between +18.5 and −18.5 degrees. Actually, I am oversimplifying here. If you take into account the absolute extremes, caused by the Earth-Moon distance variations, the Moon can range between +28.7 and −28.7 Dec, as it does in 2006, 2025, and 2043. Regardless of which hemisphere you live in, the full Moon is highest in winter, the first-quarter Moon is highest in spring, and the last-quarter Moon is highest in autumn. The new or old crescent moons, being so close to the Sun, are at their highest from late spring to early autumn.

There are two additional orbital factors that the keen lunar imager will learn to appreciate: the variation in lunar distance and the effects of libration.

Perigee, Apogee, and the Far Side

The Moon's monthly orbit around the Earth is not circular, it is elliptical. At its closest (perigee), the Moon's center is only 356,410 kilometers from the Earth's center. At its furthest (apogee), the centers are 406,697 kilometers away. That is a considerable variation of ±7%. On average, a tiny crater on the Moon, say one kilometer across, will only span an angle of around 0.55 arc-seconds and will be close to the limit of detection in average seeing conditions, with an amateur astronomical telescope. Having said this, tiny high contrast rilles, seen at low sunset and sunrise illumination angles, will still be detectable even though they may be well below a kilometer in width.

Libration is a phenomenon by which we are able to peer around the lunar limb and see, at a painfully shallow angle, the edges of the Moon's far side. In fact, we can theoretically see 59% of the lunar surface, not just 50%, by using libration tilts. Let us dispel one myth here: contrary to popular terminology there is not a lighting orientation that results in a permanent "dark side of the Moon," so viewing features on the far side is not complicated by this misconception. When the Moon appears full to us, the opposite side is dark; but when the Moon appears new (i.e., a hair-thin crescent) the opposite side is almost fully illuminated. However, the Moon does have a near side and a far side because the Moon's rotation on its axis has become "locked," due to tidal forces; thus, the lunar side that we can see stares permanently down at the Earth (if we ignore the libration effects that is). The "man in the Moon" never rotates out of view.

The Moon rotates around the Earth every 29.5 days (from new Moon to new Moon) and, with respect to the stars, every 27.3 days (remember the whole Earth-Moon system orbits the Sun, too). However, it is constantly rotating on its axis such that the same face always points towards the Earth . . . well, almost. Imagine you are talking face to face with someone, but occasionally they lower their head, so you see more of their badly fitting wig, or raise their head so you see more of their chin. Now and again they shake their head too, so you see a bit more of one ear and then the other. This is analogous to what librations do to our view of the Moon. Because the Moon's orbit is elliptical, its angular position with respect to the Earth does not vary constantly, even though its axial rotation is always the same. The velocity of the Moon around the Earth is faster at perigee than apogee. Because of this, we can sometimes peer around either the eastern or western limbs and see almost 8 degrees more Moon (7° 54' to be precise). This is called a libration in longitude.

There is an additional effect called diurnal libration, caused by the fact that the Earth has a radius of over 6,000 kilometers and so, depending on whether the Moon is rising or setting (or if you are at the north or south poles for that matter) you are standing on a platform that gives you an extra ability to peer round the limb. However, most observers will view the Moon when it is near their meridian and will not travel to the Arctic or Antarctic to get a tiny bit more north-south advantage!

The main librations in latitude (excluding traveling to the poles of the Earth) are caused by the fact that the lunar equator is tilted with respect to the lunar orbital plane, much as the Earth's equator/axis is tilted with respect to the ecliptic/ecliptic pole. Thus, as the Moon orbits the Earth, first one pole and then the other tilts by 6° 41' toward the Earth (the absolute extreme librations are actually 6° 50').

In practice these monthly librations in latitude and longitude form a vector sum, peaking in a maximum libration effect of 10 degrees when latitude and longitude librations peak together and swing a feature on the NE, SE, SW, or NW limb toward Earth. Of course, that is not much use if the feature is in darkness, but it is highly exciting when a favorable Sun angle picks out the feature well and the sky is clear.

Under extreme libration conditions the lunar webcam user can secure some rare shots of regions usually invisible from the Earth. Of course, the Moon has been fully mapped by various spacecraft, most recently by Clementine and Smart-1. Even the southern polar regions have at last now been fully imaged. However, this does not take away the enjoyment of observing and imaging the Moon. Despite the Moon being fully mapped, it is true to say that it has certainly not been mapped at high resolution at every illumination angle. Following the progress of mountain peak shadows as they grow and shrink under sunset and sunrise conditions can be a fascinating pastime and can be carried out at unprecedented resolution in the webcam era. Lunar craters can easily become like "old friends" to the dedicated lunar observer and, with webcams, it is now possible to capture the appearance of a crater without any hint of observer bias.

Unlike any other planetary body, the Moon is an extremely photogenic subject, even under poor seeing conditions, simply because it is so large. When seeing is poor, just reduce the f-ratio (remove the Barlow lens or Powermate) and shoot some images at a scale of, say, 0.5 or 1.0 arc-seconds per pixel. The rugged southern highlands and the jagged southern limb regions are especially good wide-field targets in this respect. Using a webcam on a small-aperture Newtonian, a mosaic of half a dozen frames can produce a spectacular picture of an entire lunar crescent. Alternatively, a digital camera or digital SLR can be used to good effect too.

Color, Monochrome, and Terminator Shadows

Although the Moon does show subtle hues and shades, it is, essentially, a rocky, mountainous, and airless world, without the colored features that planetary atmospheres produce. The most prominent colors seen on the lunar surface are the prismatic rainbows of color produced by dispersion in the Earth's atmosphere. Therefore, sensitive monochrome webcams like the ATiK 1HS can be used to good effect. To improve the resolution on such a bright object as the Moon, such webcams can be used in conjunction with near-infrared filters to produce stunning results. The French astronomer Bruno Daversin has produced some truly staggering lunar images using this technique with a 60-cm Cassegrain telescope. In good seeing a green filter can sharpen the view by reducing dispersion.

When imaging the planets one has to be aware that they rotate and thus a webcam video of more than a few minutes duration will smear details on the planet. Does such a time limit exist on lunar features? The answer, surprisingly, is yes!

The lunar terminator (the night/day boundary) sweeps across the lunar surface at a rate of 15 kilometers per hour at the lunar equator. For formations at latitudes nearer to the polar regions the speed is slower of course, simply because the Moon's circumference is less at higher latitudes. (Cosine of latitude is the ratio; e.g., at latitude 60° north the terminator moves at half the equatorial rate.) If you are looking at features on the lunar equator and the meridian (in other words in the dead center of the disc), the terminator speed will translate into an angular movement of about 8 arc-seconds per hour. Assuming we wish to image features as small as 0.3 arc-seconds, in the best seeing conditions, this gives us a time window of about two minutes in which to capture our webcam AVI video, if we do not want the terminator to move during our shot. This is a similar challenge to imaging the planet Jupiter. There is another factor too that is worth considering. At the very low Sun angles that exist right on the terminator, shadows of very high mountain peaks cast onto the lunar terrain can lengthen and shrink rapidly at low Sun angles. This too is a reason why imaging lunar terrain, at high resolution, over tens of minutes, will definitely lose you detail. However, in practice, a five minute window will work fine, even on the lunar center, as shadows are diffuse objects due to the angular diameter of the Sun. It is sometimes said that you never see a crater on the Moon, at sunrise/sunset conditions, looking the same. This is very true. Lunar librations can make craters near the limb change from almost circular to highly elliptical in extreme conditions. In addition, sunrise or sunset on a formation can occur later or earlier than expected. A healthy five-minute window will give you plenty of webcam frames to stack but, with the Moon, you can often be a bit more selective, as it is so bright and has a great deal of contrast. In addition, distortions over wide fields are far more obvious when imaging big craters that span an arc-minute or two in size. Stacking a few dozen excellent frames will often give a sharper picture than stacking hundreds of distorted ones. The exception is the case when we are trying to resolve tiny low-contrast craterlets on a dark crater floor: here, signal-to-noise is crucial and we need all the frames we can get. The Moon makes a spectacular picture at any f-ratio. When seeing is poor, going down to f/20, f/15, or even using a digital SLR can produce a spectacular picture. To get the whole Moon in the shot is quite feasible with a DSLR. At a focal length of 1.5 meters the Moon will be 13 mm across its diameter and so will fit onto a typical DSLR chip easily.

The Sun's Corrected Selenographic Colongitude

This next bit is a quite a brain-twister, but you do not need to understand it to image the Moon. However, it may help you understand when a critical illumination will recur, if you get bitten by the lunar imaging bug. The standard reference for finding out the altitude of the Sun above the lunar terrain (and therefore where the night/day terminator falls) is a table of the Sun's selenographic colongitude (SSC). This is a bit of a mouthful but it simply means the height of the Sun (or its longitude) as seen from the lunar surface. When the SSC is 0°, the Moon is around the first quarter phase; at 90°, the Moon is full; at 180°, it is last Quarter; and, at 270°, the phase is new. This might seem like a good way of predicting exactly when the terminator will cross a lunar formation. However, there is a fly in the ointment because of the fact that the Moon's axis of rotation can be tilted with respect to the Sun, too, by plus or minus 1.5°. For high northerly and southerly objects, a correction to the SSC is required to tell you where the terminator will precisely fall. The correction, c (in degrees) can be derived from the formula $c = \arcsin(\tan b \times \tan i)$, where c is the correction required, b is the latitude of the formation, and i is the latitude of Sun (N+, S−), otherwise known as the Sun's selenographic latitude. The correction will be + or − depending on whether the Sun and the formation are on the same or opposite sides of the lunar equator respectively. A + correction means that sunrise occurs early and sunset occurs late on the formation, both of which produce shadows that are shorter than the tabulated SSC would indicate; a − correction means just the opposite, of course. Incidentally one degree of longitude = 1.97 hours. A related trap that might conceivably cause problems when trying to see a precise illumination angle is the fact that for formations situated away from the lunar equator, the difference between the corrected SSC and the longitude of the formation is not a measure of the Sun's altitude there; the difference must be multiplied by cos *b* to obtain the true solar altitude (for low Sun angles). I am indebted to lunar expert Ewen Whitaker for explaining all this to me in a letter some 20 years ago.

Confused? Don't worry! Essentially you just need to bear in mind that, due to librations, you are never quite looking at a lunar formation from the same viewpoint and, due to the Moon's axial tilt with respect to the Sun, the terminator is rarely lying at exactly the same angle across the Moon's face. Will this make any actual difference to your observing, in practice? Well, maybe. If you want to image a lunar feature with a similar play of shadow and light as on a previous occasion, you may be a few hours early or late if you do not correct the SSC. All this is just a great reason to keep imaging the Moon. It never looks exactly the same.

Transient Lunar Phenomena (TLP)

Since the 1950s, various controversial glows and obscurations have been reported on the lunar surface by amateur astronomers. In general, these "events" have been associated with specific craters such as Alphonsus, Plato, Aristarchus, and Gassendi, although much more obscure craters, like the diminutive Torricelli B, have also generated alerts. Undoubtedly TLP "hunting" is on the very fringe of amateur astronomy, on the borderline between science and pseudoscience and, as such, it attracts a fair number of cranks. There is no shortage of weirdos out there who are convinced that alien abductions really occur or that astrology means something. There is also no shortage of cranks who want to be regarded as the next Einstein or Stephen Hawking, but without putting in the mental effort to indulge in mainstream science. For 11 years (1980–1991) I was an active member of the BAA Lunar Section TLP network: a network of lunar observers dedicated to responding quickly to claims of glows and obscurations on the Moon. I had a rather unique role in that team. I was determined to photograph, and later videotape, craters during TLP alerts, in order to try to prove or disprove what was happening. What did I gain from that experience? Well, the main thing I learned was that not only does the Moon look different every night due to librations, lunar distance, and the position of the terminator, it looks very different due to the effects of atmospheric spurious color. If you look at the crater Plato when the Moon is high in the sky, at lunar perigee (closest to Earth) and with a libration moving it nearer to the center of the disc it looks perfectly normal. However, if you look at it when low down, close to apogee and when librations make the crater look highly elliptical it looks blurred and fringed with color. It is under those latter circumstances that TLP alerts were normally generated. The imaging situation in the 21st century is far removed from the situation in the early 1980s when I was a keen lunar observer. In those days the human eye could easily see more detail than a photograph could capture. Now the situation is reversed. A stacked composite of hundreds of webcam frames can capture all the details that even the keenest observer can see, and more . . . and guess what . . .? Mysteriously, there are virtually no TLP being reported! I think this reveals TLP for what they really are: effects of the Earth's atmosphere.

However, the BAA Lunar Section is now analyzing all those old TLP reports and alerting interested parties to repeat illuminations of features that generated TLP alerts in the distant past, to try to see whether the illumination angle of the feature is the critical aspect. A list of these features is generated by Dr. Tony Cook and is available at www.lpl.arizona.edu/~rhill/alpo/lunarstuff/ltp.html.

Some of the historic TLP were reported by some quite famous names in astronomy so it would be nice to think that there was some substance to them, even if I remain rather sceptical, despite being a past member of the BAA TLP team! The red glow in the crater Gassendi, on April 30, 1966, was a particularly well-documented mystery and a handful of other cases make me wonder whether I should be such a cynic!

Some Awesome Lunar Features

Over the next few pages I will be referring to my 20 (or so) favorite craters and regions on the Moon. To help locate these on the lunar disc please refer to Figures 10.1a (the major seas) and 10.1b (the actual features) and their keys. This is not a book about the Moon, so the list is far from exhaustive. However, there are enough targets here to keep the webcam imager happy. As with all the images in this book, I have placed south at the top. This matches the visual view through a Newtonian telescope and amateur astronomers observing the Moon and planets have placed south at the top since the dawn of the telescopic era, so I will keep to that tradition.

Test Objects

A number of lunar features have become test sites for a telescope's resolution. Probably the most famous features in this regard are the tiny craterlets on the floor of the crater Plato (shown later in this chapter in Figure 10.21), the rilles near the

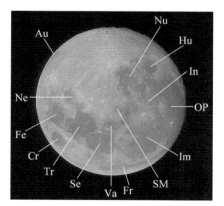

Figure 10.1a. The major lunar seas. Key (clockwise from top): Nu = Mare Nubium; Hu = Mare Humorum; In = Mare Insularum; OP = Oceanus Procellarum; Im = Mare Imbrium; SM = Sinus Medii; Fr = Mare Frigoris; Va = Mare Vaporum; Se = Mare Serenitatis; Tr = Mare Tranquillitatis; Cr = Mare Crisium; Fe = Mare Fecunditatis; Ne = Mare Nectaris; Au = Mare Australis. Original image captured by Jamie Cooper, using an Orion Optics SPX Newtonian (250 mm f/6.3) and Canon 300D.

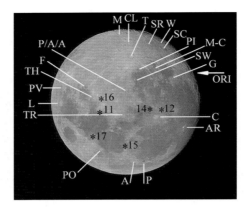

Figure 10.1b. The author's favorite webcam imaging targets referred to in the text. Craters and geological features are indicated in white and Apollo landing sites (11, 12, 14 – 17) in black. Apollo 13 did not land on the moon due to the onboard emergency. In keeping with the visual telescopic view, all images have south at the top. Key (clockwise from top): M = Moretus; CL = Clavius; T = Tycho; SR = Schiller; W = Wargentin; SC = Schickard; Pl = Pitatus; M-C = Mercator/Campanus and the Hippalus rille region; SW = Straight Wall; G = Gassendi; ORI = Mare Orientale *behind* the limb and invisible; C = Copernicus; AR = Aristarchus; P = Plato; A = Alpine Valley; PO = Posidonius; TR = Triesnecker; L = Langrenus; PV = Petavius; TH = Theophilus and Cyrillus; F = Fracastorius; P/A/A = Ptolemaus, Alphonsus & Arzachel crater chain. Original image captured by Jamie Cooper, using an Orion Optics SPX Newtonian (250 mm f/6.3) and Canon 300D.

crater Triesnecker (Figure 10.2), and the Alpine Valley rille (Figure 10.3). All of these targets are excellent ones for testing that new instrument and, unlike planetary features, they are (literally) cast in stone and available every month. The lunar crater Plato is well known to all lunar observers and is sited at 9° west and 52° north, squeezed between the narrow Mare Frigoris and the northern edge of the Mare Imbrium. Plato can appear almost circular when a southerly libration tilts it nearly 7° toward the center of the Moon's disc. But it appears highly elliptical when northerly librations tilt it 7° toward the north of the disc. Under good seeing conditions and a suitable illumination even a new observer will be able to glimpse half a dozen tiny craterlets on the smooth, dark floor of Plato. The largest of these craterlets is the central one, which has a diameter of three kilometers. The next three largest floor craterlets (two of which are side-by-side) are between two and three kilometers in diameter, but the rest are much smaller. Just how many craterlets are on the floor of Plato and, specifically, how many can be imaged from Earth? This is a very good question and one that does not have a definite answer. Using a Celestron 11 at a focal ratio of 31, and an AtiK webcam, the U.K.'s Damian Peach has resolved 22 Plato craterlets, the smallest ones being barely more than 0.5 kilometers across. This translates into an angular resolution of about 0.3 arc-seconds, or slightly below the theoretical resolution of a 28-cm (11-inch) aperture. But this by no means represents the limit of Earth-based resolution. Bruno Daversin, using a 60-cm f/16 Cassegrain at the Ludiver facility near Cherbourg in northern France (l'Observatoire Planetarium du Cap de la Hague, www.ludiver.com) has recorded 100 tiny craterlets on Plato's floor, some only 0.3 kilometers across. At these tiny

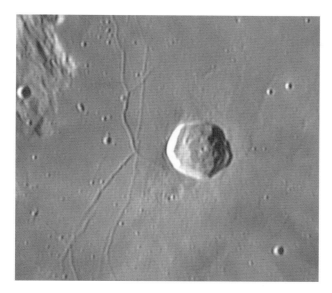

Figure 10.2. Triesnecker and its rille system imaged on September 5, 2004, with a ToUcam Pro webcam and a 250mm f/6.3 newtonian at f/38. The field spans 110 kilometers. Image: M. Mobberley.

Figure 10.3. The Alpine Valley and its rille imaged by Damian Peach on March 1st 2004 with a Celestron 11 SCT and ATiK 1HS webcam.

angles, so close to the telescope/signal-to-noise limits of the equipment, stacking as many images as possible is essential, even on such a high-contrast feature as the Moon. Tiny features only emerge when noise levels are low and noise decreases with the square root of the number of images stacked. On nights of poor seeing, especially near full Moon, even Plato's central craterlet may be hard to spot. However, under good seeing even a 25-cm telescope can reveal a dozen or more craterlets to the visual observer.

The rilles close to the small crater Triesnecker have long been a test of both optics and atmospheric stability for the lunar observer. Back in June 1981, *Sky & Telescope* featured a remarkable photograph of the Triesnecker rilles (page 512), taken by U.S. amateurs Thomas Pope and Thomas Osypowski in 1964. Using a 31.8-cm (12.5-inch) reflector at f/60 they had resolved craterlets only two or three kilometers across in the Triesnecker region. The one-second exposure was one of the most remarkable of its era and yet these days it could be equalled with a 100-mm telescope and a webcam!

Triesnecker itself is only 25 kilometers in diameter and the widest of the Triesnecker rilles are only one arc-second (less than two kilometers) across. In addition, there is an assortment of tiny craterlets all around the region, varying from three to less than 0.5 kilometers in diameter and these provide an excellent series of resolution test objects. Once again, Bruno Daversin has achieved the best image of the region (at least the best I have seen) with the 60-cm f/16 Cassegrain of the Ludiver Observatory. Incidentally, the crew of Apollo 10 took a breathtaking and historic photograph of the Triesnecker crater and rilles, from a distance of 140 kilometers, in 1969. Comparison of the region with Daversin's photograph can be most instructive.

Unlike Plato's craterlets and the Triesnecker rilles, the Alpine Valley rille is a single, highly elusive feature that only the keenest eyed amateurs had recorded visually before the webcam era. Indeed, even as recently as the 1940s, some observers doubted its existence: it is very elusive unless seeing conditions are excellent. The Alpine Valley is situated at 49°N, 3°E and is more properly called Vallis Alpes. Almost 180 kilometers in length it cuts through the Montes Alpes that separate the Mare Frigoris from the Mare Imbrium. The tiny sinuous rille on the valley floor can be glimpsed in a 15-cm telescope, although I have rarely seen it in any instrument. It barely exceeds 1.5 kilometers in width (approx 0.8 arc-seconds) along its length and runs at a 45° slope to the north-south line, so can never be as favorably illuminated as the Triesnecker rilles. In terms of a resolution target it is, perhaps, the ultimate that the aspiring lunar imager can strive for.

The Apollo Landing Sites

As someone who lived through the Apollo Moon landing era at a very impressionable age (I was 11 in 1969), I still find hunting down and imaging the sites where men walked on the Moon a fascinating challenge. I often wonder, if the Apollo missions were to be repeated in the 21st century, what the modern imagers would be able to achieve when looking at a spacecraft? Webcam users have already taken impressive images of the International Space Station in orbit, maybe they could have captured the Command and Lunar Module docking in Earth orbit, too? Or maybe they could have imaged engine burns in lunar orbit or dust thrown up from the third stage rockets that were deliberately crash-landed on the Moon?

Although astronauts have not ventured beyond Earth orbit since 1972, you can still relive those extraordinarily exciting times by locating the landing sites using a relatively modest telescope. With a 25-cm reflector and a decent map, lunar features under 1 kilometer across can be glimpsed, especially when the lunar terminator is close to the region under scrutiny and the surface relief is accentuated by deep sunrise and sunset shadows.

Apollo 11

"Houston, Tranquillity Base here. The Eagle has landed." "Roger, Tranquillity, we copy you on the ground. You've gotta bunch of guys about to turn blue. We're breathing again. Thanks a lot." This exchange between Neil Armstrong and Charlie Duke sent shivers down my spine as an 11-year-old schoolboy glued to the BBC coverage (anchored by James Burke and Patrick Moore). My 30-mm refractor was soon trained on the Moon, but (not surprisingly) did not show the region well! Eagle's landing site was in the Sea of Tranquillity, not far from the two 30-kilometer craters Sabine and Ritter. If you draw a line from the center of Sabine to the small 7-kilometer crater Moltke, at two-thirds of the way to Moltke you pass 20 kilometers south of the Apollo 11 landing site. A day or so before half Moon, i.e., the first quarter phase, the region is thrown into sharp relief and the Hypatia rille, near Moltke is well shown too. If you have a big telescope and a good map, the tiny craters Armstrong, Aldrin, and Collins can just be glimpsed at high powers. With a webcam, under good seeing, it is dead easy to record these tiny craters with a 20-cm and larger instrument. The Apollo 11 Landing Site is at 0.65° N, 23.51° E and the landing date was July 20, 1969. See Figure 10.4.

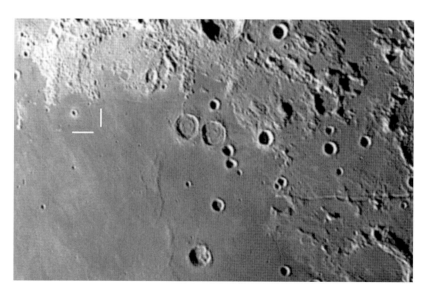

Figure 10.4. The Apollo 11 Landing site imaged by Mike Brown of York, England, with his 37-cm Newtonian at f/12. The craters Ritter and Sabine are nearby. 2x Barlow plus Starlight Xpress HX516 CCD.

Apollo 12

Three hundred fifty kilometers SSW of the spectacular crater Copernicus you will find the 40-kilometer crater Lansberg, the closest large crater to the Apollo 12 landing site on the Mare Insularum (Sea of Isles). If you look almost three Lansberg diameters to the ESE of that crater, you are staring at the landing site of the Lunar Module Intrepid on November 19, 1969. The position is at 3.04° S, 23.42° W. Roughly four or five days before full Moon the region is perfectly illuminated for intense scrutiny. Remarkably, Apollo 12 landed only 180 meters from the Surveyor 3 probe, which had landed on the lunar surface 31 months earlier. The Moon rocket was launched in very dodgy weather conditions and was struck by lightning 30 seconds after launch. After hitting the massive, 111-meter tall Saturn V rocket, the bolt travelled down the vapor trail to the launch pad! See Figure 10.5.

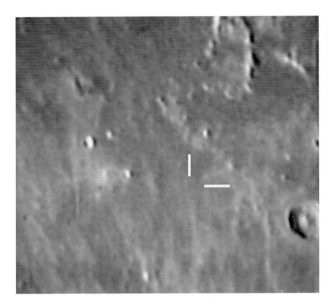

Figure 10.5. The Apollo 12 landing site. Field is 260 kilometers wide and the crater Lansberg is in the top left of the frame. Image from a CCD video by the author with a 36-cm Cassegrain in May 1987.

Apollo 14

After the near-disaster, but spectacular rescue, of the Apollo 13 mission, almost 15 months elapsed between Apollo 12 and Apollo 14. However, on February 5, 1971, the fifth and sixth NASA astronauts, Alan Shepard and Ed Mitchell landed the Lunar Module Antares in the rugged Fra Mauro region of the Moon, only 160 kilometers east of the Apollo 12 landing site. The position is 3.66° S, 17.48° W. Once again, the large crater Copernicus is the best stepping stone to finding this region through a telescope and the low Sun angle about five days prior to full Moon will throw the craggy terrain into sharp relief. Four hundred kilometers to the SSE of Copernicus you will find the remains of a circular walled plain almost 100 kilometers in diameter. This is Fra Mauro. It is crossed by distinctive rifts on the crater floor. Roughly 20 kilometers north of the rim of this plain is the Apollo 14 landing site. See Figure 10.6.

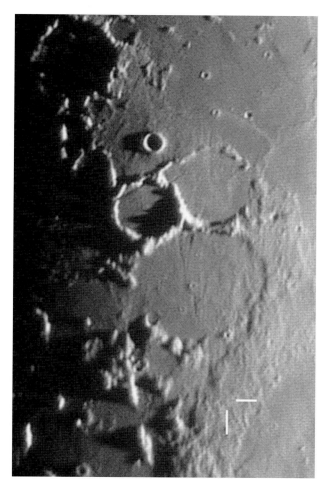

Figure 10.6. The Apollo 14 landing site imaged by Mike Brown with a 37-cm Newtonian and HX516 CCD. Image 350 × 220 kilometers. The walled plain Fra Mauro sits just north of the adjoining craters Bonpland and Parry.

Apollo 15

The first of the so-called 'J Series' missions, geared for maximum scientific gain, Apollo 15 took the first car to the Moon: the LRV or Lunar Roving Vehicle. The landing was made on July 30, 1971. At last, the astronauts could cover many miles on the lunar surface. Of all the Apollo landing sites, the Apollo 15 Hadley rille region is the most fascinating to inspect through an amateur telescope and the most infectious to webcam. Lying 200 kilometers southeast of the 80-kilometer crater Archimedes, the rille itself is clearly visible in good seeing, even though it is only half a kilometer wide. The meanderings of the rille, visited by astronauts Dave Scott and Jim Irwin, can be examined at high powers. Webcam images of the region with amateur telescopes can resolve details as small as 0.5 kilometers, much smaller than the journeys the astronauts went on in their LRV. To be able to resolve details from your back garden smaller than the astronauts rovings is quite something when you are sitting on the Earth some 380,000 kilometers away! The landing site is at 26.08° N, 3.65°E. The Lunar Module Falcon landed further from the lunar equator than any other Apollo mission and the backdrop of the surrounding hills was truly spectacular. Viewing this region a day or so after the first quarter phase, or a day or two before last quarter (if you like pre-dawn observing) will bring very rewarding views and, with a vivid imagination, you might believe that with a bit better resolution you might even glimpse the lunar module descent stage. If only! See Figure 10.7.

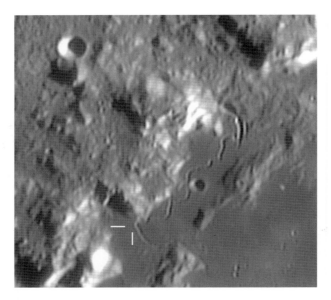

Figure 10.7. The Hadley Rille/Apollo 15 landing site imaged September 5, 2004, with a ToUcam Pro and 250-mm f/6.3 Newtonian at f/38. Image: M. Mobberley.

Apollo 16

Three hundred kilometers east of the massive adjacent craters Albategnius and Hipparchus, you will find the Cayley plains north of the crater Descartes. It was in this rugged terrain (at 8.99° S, 15.51° E) that the Apollo 16 Lunar Module, Orion, landed and astronauts John Young and Charlie Duke (Houston Capsule

Communicator on Apollo 11) went walkabout. John Young was one of only three men to have been to the Moon twice. He had also been on Apollo 10 and was the first Shuttle commander, too. Only two other astronauts went to the Moon twice, Jim Lovell on Apollos 8 and 13 and Gene Cernan on Apollos 10 and 17. However, no one *walked* on the Moon on more than one mission. The region is optimally illuminated at first quarter and, although not as distinctive as the Apollo 15 site, once the crater Descartes and the smaller crater Dollond have been located you will have the region in the eyepiece field. Apollo 16 landed on the Moon on April 20, 1972. See Figure 10.8.

Figure 10.8. The Apollo 16 landing site. The Craters Albategnius (lower) and Hipparchus are on the right hand side of this 500-kilometer wide image taken with the 74-inch Kottamia reflector in Egypt in 1965. Apollo 16 landed just north of the white area north of the crater Descartes. From an image provided by the late Dr. T.W. Rackham.

Apollo 17

The last astronauts to walk on the Moon, landing on December 11, 1972, were Eugene Cernan and Dr. Harrison (Jack) Schmidt, an astronaut-geologist. The landing site at 20.17°N, 30.77°E, the furthest from the lunar meridian, was in the Taurus-Littrow region of the Moon, just off the eastern edge of the Mare Serenitatis. Once you have located the junction of the Mare Serenitatis and the Mare Tranquillitatis and the 43-kilometer crater Plinius, you can hop to the craters Vitruvius and Littrow in the highlands bordering the region. Thirty kilometers below Littrow's southwestern edge is the landing site of the Apollo 17 Lunar Module Challenger. See Figure 10.9.

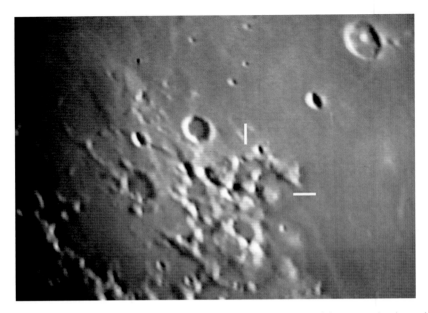

Figure 10.9. The Apollo 17 landing site from a CCD video image of the region by the author in December 1985, using a 36-cm Newtonian. The field is 400 kilometers wide and the crater Plinius is in the top right-hand corner.

The Great Lunar Craters

Every part of the lunar surface is fascinating to examine visually and to record with a webcam, but I would be lying if I did not say that there are some 20 or so craters that are favorite objects among amateur astronomers. These craters really make you appreciate how lucky we are to have such a nearby planet to examine, month after month. After a year or so of observing and imaging the Moon, these craters will become etched in your memory and you will look forward to those critical interplays of light and shadow on the craters' floors. You will wonder when you will next see a crater at a specific illumination and just how far the shadow from this or that mountain peak can stretch across the crater floor. You will dream of a clear night and steady seeing when you can just capture that critical phase that you have never seen before. One book, now out of print, that really captures the essence of addictive lunar observing is highly recommended. It is Harold Hill's *A Portfolio of Lunar Drawings*, published by Cambridge University Press in 1991. Hill (not to be confused with the stand-up comedian of the same name!) was, perhaps, the greatest lunar observer and artist of all time and his exquisite sketches, made over a lifetime of observing, still inspire me, even though I could not sketch a crater to save my life!

This is not a book about the Moon, but about using webcams. However, I think I would be neglecting my duties if I did not direct the beginner to some of the Moon's most spectacular craters. My list is by no means comprehensive; it largely just consists of my favorite craters and regions. For a full account of how to observe the Moon I would recommend acquiring a copy of Gerald North's excellent book *Observing the Moon*, which is lavishly illustrated with photographs from the 1.55-meter Catalina telescope in Arizona. Taken in the 1960s, these photographs are of a similar resolution to modern amateur images.

Here are some of the Moon's best features (see Figure 10.1b. for their location).

Moretus [70.6° S, 5.5° W]

Many keen amateur astronomers will not be familiar with my first choice target for lunar imaging, but the 114-kilometer crater Moretus is an absolutely stunning crater. Lying just west of the central meridian and among the rugged lunar highlands, this crater has some of the most dramatic wall terracing of any crater on the Moon. Under sunrise illumination and good seeing, the inner-wall detail is simply awesome as perfectly captured by Damian Peach in Figure 10.10.

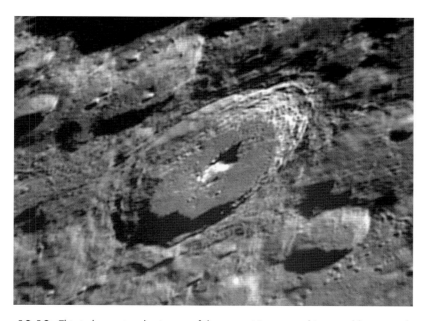

Figure 10.10. This truly spectacular image of the crater Moretus and its superbly terraced walls was captured using a Celestron 9.25 Schmidt-Cassegrain and an ATiK 1HS webcam on March 20, 2004. The telescope was working at f/40. Image: Damian Peach.

Clavius [58.4° S, 14.4° W]

It is tempting to think of Clavius (Figure 10.11) as the largest crater on the near side of the Moon. In fact, this is not the case. But, undoubtedly, it is the largest spectacular crater on the lunar surface and it is a truly awesome sight. Larger,

Figure 10.11. An excellent image of the huge crater Clavius, taken with a 250-mm f/6.3 Orion Optics Newtonian, working at f/24. Image taken with an ATiK 1HS webcam on March 19, 2004. Image: Jamie Cooper.

much older craters, do exist but are harder to spot as a complete crater; they appear as a "ruined plain," ruined by impacts after they formed. The largest and deepest basin on the Moon is, in fact, the massive southern polar Aitken Basin, which is 2,500 kilometers deep and 12 kilometers below the mean sphere! However, from the viewpoint of Earth-based visual and webcam observers, the ruined crater Bailly, with a diameter of 287 kilometers (situated at 66.5° S, 69.1° W) is the largest named crater, but is so foreshortened that it is in no way comparable with Clavius. Clavius itself lies at 58.8° S and 14.1° W in the rugged southern highlands and is an awesome 245 kilometers in diameter. It is only 200 kilometers northwest of Moretus. The crater first becomes illuminated a day or so after first quarter and the Sun does not set on the formation until after last quarter. This factor, coupled with its sheer size, gives Clavius the illusion of being an almost permanent feature through the eyepiece. There are regions of the Clavius floor that have, for decades, been a test for the lunar photographer and are now tests for the small-aperture webcam user. I can remember sitting in the legendary Horace Dall's study in Luton in 1984, when he proudly showed me his best photographs of a horseshoe-like pattern of small craterlets on Clavius' floor. The craterlets, only 2 kilometers across, were at the very limit of resolution in the film era (even with Dall's 39-cm Dall-Kirkham Cassegrain). Recently, Damian Peach showed me a webcam composite he had taken with an 80-mm apochromat that was every bit as good as Dall's image! The webcam really has opened our eyes to what is possible.

Clavius is so large that there are numerous decent sized crater's within its walls! The most distinctive is the 50-kilometer crater Rutherfurd and the crescent-shaped curve of diminishing crater's (Clavius D, C, N, and J) which lead off from it. Another 50-kilometer crater called Porter breaks into Clavius' northeastern wall. The only trouble with Clavius is fitting it onto the webcam chip. If you are a planetary imager who frequently increases the telescope's effective focal length to almost 10 meters you may have to review your strategy. Clavius spans over two arc-minutes in length and so an effective focal length of five meters or less is needed to comfortably squeeze it onto a webcam's CCD chip.

Tycho [43.3° S, 11.2° W]

A deep crater (4.85 kilometers) and youthful by lunar standards (roughly 100 million years old), Tycho (Figure 10.12) is another spectacular feature that dominates its surroundings. However, this 85-kilometer crater is sited in the rugged southern uplands where it can almost be lost at very low Sun angles when shadows of mountains and craters confuse the scene. In fact, Tycho is one of the few craters that dominate at full Moon. Under these conditions, with no shadows, the Moon is a rather bland and dazzling feature. However, Tycho's spectacular ray system dominates much of the southern hemisphere at full Moon and the crater itself becomes a glowing white ring, with a dark halo and a bright center. Of course, Tycho's rays

Figure 10.12. The crater Tycho imaged with a small 80-mm Vixen apochromatic refractor working at f/45. ATiK 1HS webcam. Image: Damian Peach.

are this prominent because it is such a relatively new impact crater and other smaller impacts, from asteroids, meteoroids, and micrometeoroids, have not eroded the rays. Many astro novices, when they see the full Moon through a telescope, assume that Tycho marks the lunar south pole as, subconsciously, that ray center looks like it must have a geographical significance. In fact, Tycho is only at 43 degrees south; that is a full 47 degrees north of the lunar south pole.

Schickard [44.4° S, 54.6° W], Wargentin [49.6° S, 60.2° W], and Schiller [51.8° S, 40.0° W]

I have grouped Schickard and Wargentin (Figure 10.13) and Schiller (Figure 10.14) together as they are all within a 6 arc-minute diameter circle near the SSW limb of the Moon and all of them are large and fascinating craters. Being so far from the center of the lunar disc, as viewed from Earth, all these craters appear highly foreshortened. Schickard is a monstrous crater: at 227 kilometers in diameter it is not much smaller than Clavius. However, the crater floor is far less detailed and appears relatively smooth, with two shades of "lunar gray" to it: gray and darker

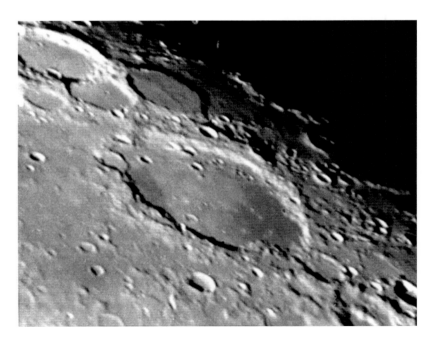

Figure 10.13. A photograph taken on February14, 1984, with Ilford XP1 400 film and a 36-cm Cassegrain working at f/70. The huge crater Schickard dominates this region with Nasmyth and Phocylides also shown. Above Schickard the lava-filled crater Wargentin is visible. Photograph: M. Mobberley.

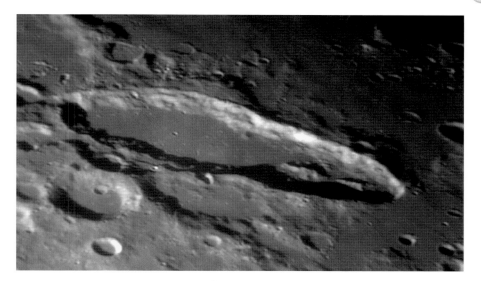

Figure 10.14. The long, thin crater Schiller. An extraordinary image taken on April 21, 2005, from Barbados, with a Lumenera video camera and a 235mm Celestron 9.25. Image: Damian Peach.

gray! However, the smooth dark floor makes it a highly distinctive feature. To the south of Schickard is a most unusual crater. In fact it does not resemble a crater at all, because the interior has been filled to the very brim with lava at some stage, making Wargentin look more like a raised smooth coin, or lozenge, stuck to the lunar surface. Wargentin is 84 kilometers across and the flat top is not totally featureless as it has a few "wrinkle ridges" radiating from the center. Wargentin itself butts up to the large overlapping craters Phocylides and Nasmyth. A few hundred kilometers to the southeast of Wargentin is one of the most distinctive large craters on the lunar surface. The crater Schiller is a highly elongated crater measuring 179 × 71 kilometres. Its long, thin shape is exaggerated further by its proximity to the lunar limb. At its northwestern end the crater floor has some distinctive mountains, although the rest of the floor looks relatively flat. Personally, I always think Schiller sounds like a long thin name and it's a name that suits this formation perfectly! The figure of Schiller by Damian Peach must be one of the finest earth-based pictures ever taken.

Pitatus [30° S, 14° W] and the Straight Wall area [22° S, 7° W]

Pitatus (Figure 10.15) is a large crater with a flooded floor on the southern edge of the Mare Nubium. The walls have been badly eroded in places, especially to the north, and it is joined by Hesiodus on its western rim. Pitatus is 100 kilometers in diameter and has a relatively small central mountain peak noticeably offset to the

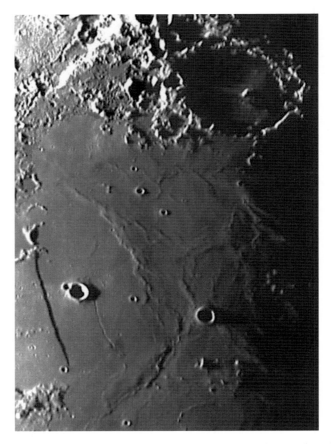

Figure 10.15. The region from the crater Pitatus (deep in shadow) to the straight wall (Rupes Recta) and the crater Birt. 37cm Newtonian with a 2x Barlow and Starlight Xpress HX516 CCD. Image: Mike Brown.

west of the crater. Although Pitatus' floor is flooded, there are some delicate rilles there, but these are very subtle features and not comparable to the spectacular rille networks found, for example, in Posidonius and Gassendi. Two hundred kilometers northeast of Pitatus' outer wall you will find the crater Birt and the feature known as Rupes Recta, or the Straight Wall (Figure 10.16), or Straight Fault. Whenever I think of this feature, I recall numerous talks I attended by Patrick Moore when he would say "It's called the Straight Wall because it's not straight and it's not a wall." This always got a good laugh from the audience! Patrick was right, but it is one of the most noticeable linear features on the Moon, even if it does have a slight curve to it. It is a geological fault that has a slope of about seven degrees, but this is enough to make it look like a wall under a low Sun angle. Although relatively small, at 17 kilometers in diameter, the crater Birt is a crater worthy of a few moments study, too. It is an almost perfect small circle, apart from where it joins the smaller crater Birt A and Rima Birt, a 50-kilometer-long rille lies to its northwest, joining the tiny craters Birt F and Birt E.

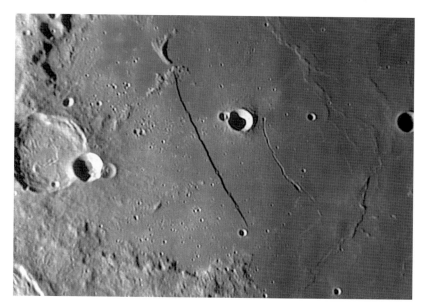

Figure 10.16. An extraordinary high-resolution shot of the straight wall, crater Birt (center) and the overlapping craters Thebit, Thebit A, and Thebit L (left-hand side). 250-mm f/6.3 Newtonian and ATiK 1HS webcam. Image: Jamie Cooper.

Mercator [29.3° S, 26.1° W], Campanus [28.0° S, 27.8° W], and the Hippalus rilles

Three hundred kilometers southeast of Gassendi you will find a truly magnificent gem on the Moon: a region that looks like a giant cat had frantically scratched the lunar surface with its claws (to the lower right of Figure 10.17). This is the region east of the ruined and flooded crater Hippalus and its associated rilles. The rilles are northwest of the striking crater pair Mercator and Campanus. There is so much to describe about this complex region that one could almost write a chapter on it alone. Fortunately, a picture speaks a thousand words and so I am spared from trying to describe the scene. The whole area straddles the boundary between the Mare Nubium and the Mare Humorum and though the Hippalus rilles are the most eye-catching feature there is plenty more to interest the visual observer or the webcam imager. The 50-kilometer craters Mercator and Campanus appear, under average seeing, to have fairly smooth, flooded floors (except for the rubble and the big craterlet on the floor of Campanus). Mercator's floor in particular looks pretty smooth and featureless. But catch a few dozen good webcam frames on a night of excellent seeing and even this crater's floor reveals tiny craterlets. Nearby the flooded crater Kies will catch your eye and further north the magnificent 60-kilometer Bullialdus is another favorite with a textbook central peak and terraced walls.

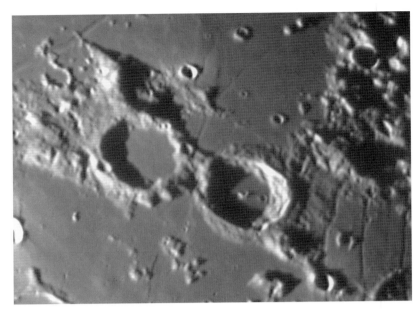

Figure 10.17. The craters Mercator (above and left of center), Campanus (below and right of center), and the Hippalus rilles (lower right). 30-cm Schmidt-Cassegrain at f/22 & ToUcam Pro webcam. A stack of 178 frames. March 1, 2004. Image: M. Mobberley.

Gassendi [17.5° S, 39.9° W]

Like Theophilus, mentioned later, Gassendi (Figure 10.18) is a distinctive crater sandwiched between a lunar sea and more rugged terrain. In this case the sandwich is between the Mare Humorum and the rugged terrain separating Humorum from the Oceanus Procellarum. Gassendi is an impressive size: roughly 110 kilometers in diameter. However, what makes it a truly outstanding object is the detail on the crater floor. Gassendi is criss-crossed by an intricate network of rilles that only truly reveal their complexity on nights of excellent seeing. There are various hills, peaks, and craterlets, too, that cast a fascinating series of shadows under critical sunrise and sunset illuminations. The significant crater Gassendi A (33 kilometers in diameter) breaks across Gassendi's northern wall as an added feature. Gassendi was the site of a major TLP (transient lunar phenomena) alert on April 30, 1966, when a number of well-known amateur astronomers observed a wedge-shaped orange/red streak extending from the wall of the crater and across the central peaks. The event has never been fully explained. Sunrise on Gassendi occurs roughly three days after the first quarter phase.

Copernicus [9.7° N, 20.0° W]

Copernicus (Figure 10.19), situated in the Oceanus Procellarum, must be the most obvious crater on the Moon to the binocular user. Although it is certainly not the largest lunar crater, it is situated close to the center of the disc and away from any

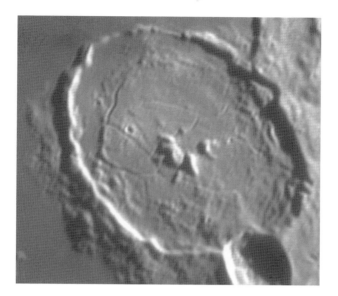

Figure 10.18. The crater Gassendi and the intricate rilles on its surface. Orion Optics 250-mm f/6.3 Newtonian and 5x TeleVue Powermate. September 9, 2004. Image: M. Mobberley.

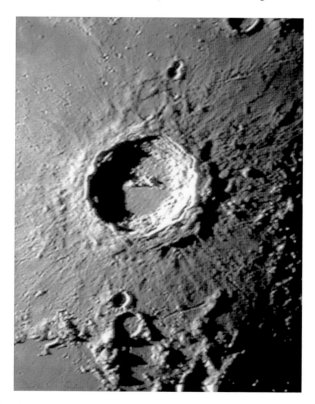

Figure 10.19. The crater Copernicus on March 30, 2004, imaged with a 20-cm Newtonian working at f/18 and a ToUcam Pro webcam. Only 22 frames were stacked for this result. Image: Mike Brown

rugged highlands. On a waxing gibbous Moon the feature is the most obvious crater a few days after the first quarter phase. The walls of Copernicus are truly magnificent and, to me, it is the only lunar crater that looks just like the top of a volcano. Of course, it is not, but it is a deep and impressive impact crater. The walls rise to 4 kilometers above the inner floor and the diameter across the crater rim is 90 kilometers. When the illumination is just right the crater walls can appear illuminated with the whole interior still in shadow, resembling a jet-black lake. Because of the rubble on the inner crater floor, below the wall ramparts, the internal floor diameter is little more than 65 kilometers. With the lunar terminator a few degrees away, i.e., just after sunrise, or just before sunset, the crater looks as deep as a bucket, but this, of course, is an illusion. One of the first instances of near-perfect atmospheric seeing that I witnessed was on October 21, 1981. It was just before dawn and a last quarter Moon was high in the sky. The view of Copernicus' floor was stunning; far more detailed than I had ever seen it. In fact, the view was so stunning that, for one crazy moment, I thought of knocking on all the neighbours' doors and dragging them from their beds for a look. (Somehow, as the temperature was close to freezing, I doubt they would have shared my enthusiasm!) It was obvious on that night, and quite a few since, just how nonsmooth the floor of Copernicus is. The terraced walls were spectacular, as was the fine detail on the floor. The renowned lunar observer T.G. Elger described Copernicus as "The Monarch of the Moon": an apt description. The outer slopes of Copernicus, in fact, the whole region within 200 kilometers of the crater center, shows the result of the impact that took place some 800 million years ago. The radial pattern looks like that of a stone thrown into mud and there are small, secondary impact craters everywhere. A good test for a lunar photograph used to be resolving the chain of "Stadius craterlets," near to the ghost ring Stadius and the magnificent, but smaller crater, Eratosthenes. However, in the webcam era, resolving the chain of secondary impact craters midway between Copernicus and Eratosthenes is not too difficult, even with a modest telescope.

Aristarchus [23.7° N, 47.4° W]

Even further west than Gassendi, Aristarchus (Figure 10.20) is an extraordinarily bright crater on the Oceanus Procellarum. The crater is so bright that the subtle banding on the crater walls is often badly overexposed. Aristarchus is only 40 kilometers in diameter but its environs are fascinating, too. Just west of this brilliant crater is the smooth-floored and relatively dull Herodotus. Meandering like a river, north and west of Herodotus is Schroter's Valley and the so-called "Cobra's head." The valley narrows from a maximum width of 10 kilometers down to 1 kilometer. The whole region is completely different from anywhere else on the lunar surface but you do have to wait until a few days before full Moon to see it emerge from the morning terminator.

Plato [51.6° N, 9.3° W]

We have already looked at Plato's smooth, dark floor with regard to resolving the tiniest craterlets, but Plato (Figure 10.21) is a fascinating crater in its own right. With such a smooth floor and such high mountain peaks on the eastern

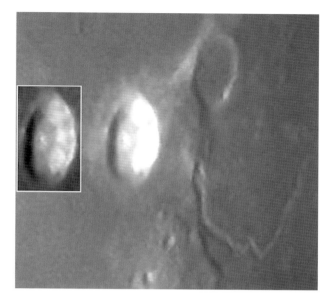

Figure 10.20. Brilliant Aristarchus, plus Herodotus and Schroter's valley, imaged with a 37-cm Newtonian working at f/14. Starlight Xpress HX516 CCD camera. A single 0.01-second exposure taken on March 29, 1999. The inset shows how the subtle banding on the western wall can be revealed with a short exposure. Image: Mike Brown.

crater rim, sunrise over Plato, at selenographic colongitudes of 10° to 13° can be a fascinating time. The highest peak on the eastern wall (just south of east) casts an incredibly straight pencil-like shadow onto the crater floor. If you catch the crater one-quarter filled with shadow, as sunrise is well underway, you will forever wonder if this needle-like shadow is capable of reaching the western wall

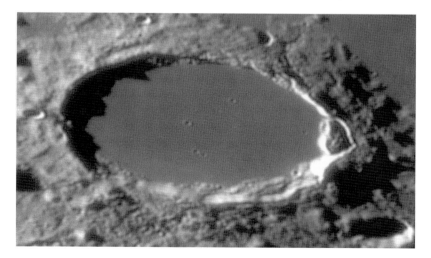

Figure 10.21. Plato and its tiny floor craterlets, imaged with an Orion Optics 250-mm f/6.3 Newtonian and an ATiK 1HS webcam on March 19, 2005. Image: Jamie Cooper.

and you will just have to resolve to catch it earlier the next month. In fact, the shadow really does reach the base of the western wall, although at the point where it just becomes visible the lighting on the crater floor is so dark anyway, compared to the dazzle from the illuminated lunar surface, that enhanced photographs or images are needed to be really sure. (Unless, that is, you are a really experienced visual observer.) In fact, the two-kilometer-high mountain peak shadow stretches an incredible 80 kilometers across Plato's floor, despite the Moon's curvature.

With webcam images and a GIF animation facility (available with most image processing packages) a dozen frames of a crater at sunrise and sunset, taken over a period of several hours, can be most instructive. An animated GIF can produce a fascinating jerky movie of shadows growing or receding. The Plato sunrise shadows are especially attractive to image, as they occur at first quarter when the Moon transits in the early evening. In spring, this is when the Moon is high in the sky.

Not far from Plato one can find the Alpine Valley and its rille, referred to earlier in the text.

Posidonius [31.8° N, 29.9° E]

Posidonius (Figure 10.22) is one of this author's favorite lunar craters. It has many similarities with Gassendi in that it fits into a corner where a lunar sea (Mare Serenitatis) borders more rugged highland terrain (north of the Taurus

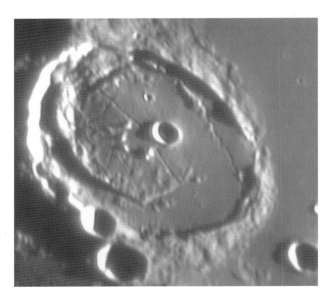

Figure 10.22. The crater Posidonius imaged with a 250-mm f/6.3 Newtonian at f/38 and a 5x TeleVue Powermate, giving a final f-ratio of 38. September 4, 2004. M. Mobberley.

Mountains). However, it lies diagonally opposite to Gassendi on the face of the Moon, as Posidonius lives in the northeast corner and Gassendi lives in the southwest corner. Posidonius is 95 kilometers in diameter and its system of rilles is just as superb as its rival on the other side of the lunar disc. Like Gassendi, the southeastern wall of Posidonius butts up against a significant crater: the 50-kilometer crater Chacornac. In addition, the floor of Posidonius contains at least one decent crater (Posidonius A) and numerous hills and peaks among the network of rilles. Under excellent seeing the rilles on Posidonius' floor are fascinating to behold. Posidonius is best placed a few days after full Moon when the evening terminator approaches the region and throws the rilles into sharp relief.

Langrenus [8.9° S, 60.9° E] and Petavius [25.3° S, 60.4° E]

It is easy for the complete novice to mistake Langrenus (Figure 10.23) for Petavius (Figure 10.24) and vice versa. Like Petavius, Langrenus lies 60 degrees east of the lunar meridian, but it is 500 kilometers further north. Langrenus is a huge (132 kilometers) crater with fabulously terraced walls, seen at an oblique angle. However, unlike Petavius there is no prominent rille stretching across from center to edge.

Figure 10.23. The crater Langrenus imaged with a modest 80-mm Vixen apochromatic refractor at f/45 and an ATiK 1HS webcam. Image: Damian Peach.

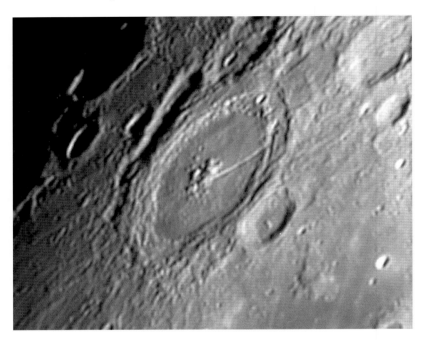

Figure 10.24. The crater Petavius imaged with a modest 80-mm Vixen apochromatic refractor at f/45 and an ATiK 1HS webcam on Sept. 1, 2004. Image: Damian Peach.

Lying at the very tip of the Mare Fecunditatis (Sea of Fertility), close to the southeastern limb of the visible lunar disc is the other crater, Petavius. This is a magnificent formation despite the fact it is foreshortened into an ellipse by its proximity to the lunar limb. In fact, this foreshortening gives the crater a distinctive three-dimensional appearance at high powers, as the terracing on the inner eastern wall is more obviously seen; it is almost as if you were flying in lunar orbit toward the feature. Petavius is 177 kilometers in diameter, thus dwarfing even magnificent craters like Copernicus and Tycho, and it features a dramatic set of central mountain peaks. There are a number of clefts on the crater floor but the most obvious is the distinctive rille that radiates from the central mountains and heads southwest, making it all the way to the western crater wall. This rille reminds me of the arm of a centrifuge every time I see it!

Theophilus [11.4° S, 26.4° E]

Theophilus (Figure 10.25) is one of the most photogenic craters on the Moon; not least because it forms a spectacular trio with the overlapping Cyrillus and, further south, Catharina. Theophilus is the youngest crater of the trio and, because of this, its features are the sharpest and least eroded. The triple (or quadruple) cluster of two-kilometer-high central mountain peaks is especially eye-catching. Perhaps the feature is especially spectacular because the whole trio is sandwiched between the

Figure 10.25. Theophilus, Cyrillus, and Catharina imaged in the dawn twilight on October 7th 2001. 37-cm Newtonian + Barlow. A single 0.03-sec and exposure at f/10 with a Starlight Xpress HX 516 CCD. Image: Mike Brown.

smooth Mare Nectaris to the east and the contrasting, rugged, Rupes Altai to the west. Theophilus is 26 degrees east of the Moon's central meridian and thus is best illuminated when the Moon is about four or five days past full or roughly six days after new. The Moon is usually available at a higher altitude in a dark sky in the former case but at a very sociable early-evening time in the latter case. Of course, seasonal effects mean that the first-quarter Moon is better placed in the evening in spring, whereas the last-quarter Moon is best placed in the morning skies in autumn. Because of this, most observers' best view of Theophilus will be in the spring evening sky.

Fracastorius [21.2° S, 33.0° E]

Not far from Theophilus and Cyrillus, on the southern tip of the Mare Nectaris, lies the distinctive walled plain Fracastorius (Figure 10.26). The northern crater wall is missing so that the southern edge of the Mare Nectaris seems to flood into the walled bay. Fracastorius has a diameter of 124 kilometers. Although there are

Figure 10.26. Fracastorius imaged in the early hours of September 10, 1998. 37-cm Newtonian + Barlow. 0.05sec exposure. Starlight Xpress MX5c CCD. Image: Mike Brown.

no major central mountain peaks on the crater floor, there are a myriad of tiny craters within the bay of Fracastorius as well as an ultrathin rille, which is a test of seeing and telescope even in the webcam era. Less than 100 kilometers to the northwest, the nearby crater Beaumont, just under half the size of Fracastorius, looks like a scale model of the larger crater.

Ptolemaeus [9.2° S, 1.8° W], Alphonsus [13.4° S, 2.8° W], and Arzachel [18.2° S, 1.9° W]

Lying exactly on the lunar meridian, but trailing south from just below the center of the disc, one will find three impressive craters named Ptolemaeus, Alphonsus, and Arzachel (Figure 10.27). The trio is so distinctive that it is almost impossible, even for a beginner, not to recognize these three huge craters. Ptolemaeus is the

Figure 10.27. Ptolemaeus, Alphonsus, and Arzachel. Imaged on April 9th 2002, with a 37-cm Newtonian at f/14 a ToUcam Pro Webcam. Only six frames were stacked for this composite. Image: Mike Brown.

largest, with a diameter of 153 kilometers. Being so close to the center of the lunar disc, it is arguably the largest crater that is perfectly circular to the telescopic observer. Although the dark floor is smooth in appearance, a careful inspection under low illumination reveals that it is covered with various shallow depressions and craterlets. Resolving tiny craterlets on the floors of Ptolemaeus was a challenge for the legendary U.K. telescope maker and photographer Horace Dall (1901–1986) as there are myriads of them between one and two kilometers across: the practical resolution limit in the photographic era. There is one, larger, flooded crater on Ptolemaeus floor, too. This feature is known as Ptolemaeus A or Ammonius and is nine kilometers in diameter. A ghost crater (i.e., submerged under a lava flow) labeled "B" lies immediately north of Ammonius. The giant crater Albategnius lies just to the east of the region and is a spectacular formation in its own right.

The second crater in the adjoining trio is the crater Alphonsus. Although Alphonsus is much smaller, at 119 kilometers in diameter, it is no less interesting and has some controversial history associated with it. In 1955 the astronomer

Dinsmore Alter took some photographs of Alphonsus with the 1.5-meter (60-inch) reflector at Mt. Wilson Observatory in California. He took filtered photographs in blue-violet and in infrared light and part of the floor of Alphonsus appeared blurred in the blue-violet pictures but sharp in the infrared. This is not a surprise to any webcam imager as seeing is always more stable and less scattered in the infrared. However, the results were interpreted as there being a lunar atmosphere on the floor of Alphonsus! Worse madness was to follow. On November 3, 1958, the Soviet astronomer Kozyrev claimed he had obtained a spectrograph proving that hot carbon gas was being emitted from Alphonsus at a temperature of 2,000°C. Needless to say, this prompted much debate and criticism and there has never been any modern evidence to support the claim. Indeed, many modern astronomers, amateur and professionals, have concluded that the late Mr. Kozyrev was "not sailing with a full crew," was "knitting with one needle," or was "two bricks short of a full load." However, it did reserve Alphonsus and Kozyrev a place in history, albeit a dubious one! The space probe Ranger 9 landed not far from Alphonsus' central peak on March 24, 1965, and not far from the northern tip of the westernmost of the two rilles that meander around the eastern side of Alphonsus' floor.

The third crater in the distinctive trio is the 97-kilometer-diameter crater Arzachel. Slightly smaller than Alphonsus, Arzachel is the most rugged in appearance of the trio and features some spectacularly terraced walls four kilometers in height, a substantial central peak, a rille on the eastern part of the floor, and even an internal 10 kilometer craterlet. Arzachel may be the smallest and youngest crater of the trio but, to me, it is the most interesting. An exceptionally fine webcam shot of Arzachel, by Jamie Cooper, is shown in Figure 10.28.

The Lunar Limb

On a final note, before we leave lunar orbit, I would like to say a few words about observing the lunar limb regions. It is often said that there is little point observing the Moon when it is full, simply because there are no long shadows and the finest details are always glimpsed when the Sun is very low over the formation being studied. Thus, at sunrise and sunset, maximum contrast is seen. At full Moon, the regions on virtually all the visible lunar disc are experiencing the Sun well above the lunar horizon. However, for features at the lunar limb, the opposite is the case. Close to full Moon these limb regions are experiencing sunrise or sunset conditions and this is the ideal time to bag some images showing just how rugged the lunar limb really is. Of course, the features will look incredibly foreshortened, but they will also give a far greater impression of being in a low orbit lunar spacecraft, skimming over the surface and seeing the curvature of the Moon. Some of the regions on the Moon's southern limb are incredibly rugged and, if you catch a favorable illumination and libration, the amount by which the Moon's limb deviates from a true circle can be quite staggering to behold, making the Moon resemble the rocky world it really is, rather than a perfect sphere with a few craters on it. For those who remember the 1970s BBC creatures "The Clangers" (knitted, woollen Moon creatures who lived on the Moon with a soup dragon!), the limb of the Moon always reminds me of the "Clanger" Moon where it is especially rugged.

Figure 10.28. An incredibly high-resolution image of Arzachel taken with a 250-mm f/6.3 Orion Optics Newtonian and an ATik 1HS webcam at f/38. Image captured on March 18, 2005. Image: Jamie Cooper.

Perhaps the most spectacular event is when a bright star grazes the rugged lunar limb causing the star to wink on and off behind the lunar mountains. Such events can easily be recorded with a webcam, especially when the star is of naked-eye brightness.

Of course, given a very favorable lunar libration *and* a favorable illumination, a rare glimpse of features normally regarded as being on the far side of the Moon can be secured. Perhaps the most famous feature in this category is the Mare Orientale, the center of which is situated at 20° S, 95° W. This mare is the bullseye at the center of a huge multi-ring basin. Surrounding the Mare Orientale is a giant circle of mountains called Montes Rook. At their eastern edge these mountains reach round to the near side, to 85° west in fact. But there is yet another ring of mountains surrounding the Mare Orientale at an even greater radius: the Montes Cordillera. These reach round to within 80° west longitude, i.e., 10° onto the near side at 20° south. With lunar librations capable of tilting the Moon by up to 10°, it is clear that the Cordillera Mountains and some way beyond can, indeed, be glimpsed from Earth, and under extreme conditions the Mare Orientale itself can just be glimpsed. Locating the region takes a bit of familiarity with the lunar surface though. The best marker to the area on this side is the dark, flooded basin Grimaldi located at 69° west and 5° south. Move about three Grimaldi lengths south from Grimaldi itself and then go to the nearest part of the lunar limb and that is where you will see the Cordillera Mountains and more, if a favorable libration is in place. How do you find out when a favorable libration will happen? Well, most decent planetarium PC packages, like The Sky or Guide 8.0, display the lunar libration amount and position angle (north = 0°; east = 90°; south = 180°; west = 270°) in the lunar information text that comes up when the Moon is selected or clicked on. Failing that, the most comprehensive astronomy data books, like the *Handbook of the British Astronomical Association*, list the extreme libration values every two weeks.

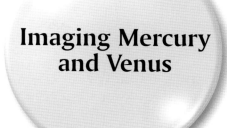

CHAPTER ELEVEN

Imaging Mercury and Venus

At first glance it might seem that there is little to be gained from imaging the two inner planets. Apart from the fact that they are only ever at a decent altitude in twilight, there would appear to be little to actually image. In addition, when at their closest to Earth they both become tiny crescents with most of their globes in total darkness. Needless to say, when imaging Mercury or Venus, great care must be taken to ensure that the Sun does not enter the telescope field. Venus is not difficult to find in broad daylight, if you know where to look, but this can be a risky business. The inexperienced amateur should never point a telescope anywhere near the Sun until a full appreciation of the risks and a great deal of experience is acquired. It is all too easy to damage one's eyesight and, unlike a webcam, an eye cannot be replaced. For information on safe imaging of the Sun and of solar transits of Mercury and Venus, please see the chapter on solar imaging. On this cautionary theme, as soon as the Sun has set, Mercury and Venus can be safely imaged without any worries about eye injuries. Atmospheric seeing conditions are usually very poor in daytime, due to solar heating. They are usually very poor an hour or two after sunset, too, as the Earth's atmosphere cools. However, during the first hour after sunset a stable period sometimes exists. This can be a good time to take high-resolution images. A declination circle or a reliable "go to" telescope are invaluable in these situations, where planets need to be found in a bright, twilight sky.

Rocky Mercury

Tiny Mercury is an airless rocky world not dissimilar in appearance to our own Moon. At 4,878 kilometers in diameter, it is 40% larger than our nearest neighbour but still smaller than Jupiter's largest Moon, Ganymede, and Saturn's largest Moon, Titan. Jupiter's Moon Callisto, at 4,806 kilometers is its nearest match.

Mercury orbits the Sun every 88 days, but rotates on its axis every 58.6 days, a bizarre sidereal day, two-thirds of the year in duration. It overtakes the Earth (on the inside track so to speak) every 116 days, at which time it can come within 80 million kilometers of the Earth and appear 12.9 arc-seconds in diameter. Of course, when this happens it is invariably only visible as a hair-thin crescent too close to the Sun for observation. Thirteen or fourteen times per century, as it nears the Earth, it actually crosses the face of the Sun (a transit), as occurred on May 7, 2003. The best time to observe Mercury is when it is at its greatest angular distance, or elongation, from the Sun. At such times the phase will be close to 50% (so it will look like a half Moon), its diameter will be 7 or 8 arc-seconds and it will be between 18 and 27 degrees from the Sun. Yes, it really does stay close to the Sun in the sky. Unfortunately, Mercury's angular elongation is not the only consideration. As well as having a good elongation, Mercury needs to be well above the horizon after sunset/before sunrise to be observed easily. Like the Sun and Moon, Mercury's declination varies throughout the year, because of the Earth's axial tilt of $23\frac{1}{2}°$. Not surprisingly, it sticks close to the Sun, but to be at a reasonable altitude in twilight at a good elongation it needs to be spring in the northern hemisphere, for an eastern elongation/evening apparition, or autumn in the northern hemisphere, for a western elongation/morning apparition. Why? Well it is all because the Earth's axial tilt will be pointing favorably to raise Mercury's altitude above the twilight horizon. Put another way, the angle the ecliptic makes with the twilight horizon is favorable. For the southern hemisphere the same rule applies, except remember that spring and autumn are six calendar months away from the corresponding northern hemisphere seasons.

Let's have another look at Mercury's bizarre day and how it rotates with respect to the Earth. What features might the webcam user be able to image? As we have seen, Mercury's sidereal (with respect to the stars) rotation period is 58.6 days. Also, it has a highly eccentric orbit around the Sun: 45.9 million kilometers at perihelion and 69.7 million kilometers at aphelion. From the viewpoint of an inhabitant of Mercury (not that anyone would want to live on a world capable of peak daytime temperatures of 430° Celsius) this creates an extraordinary situation. Because the planet moves fastest in its orbit at perihelion and the Mercurian year and sidereal day are of a similar size, the angular rotation speed of the planet in its orbit can exceed the planet's axial rotation. This means that for an observer sweltering in the daytime Mercurian heat, the Sun would actually stop moving from east to west and crawl backwards in the sky for eight days before resuming its westward travel!

As far as the earthbound observer is concerned, the northern and southern hemisphere observers tend to see the same features on the planet over many apparitions. This led the early observers of Mercury to imagine that the planet was not rotating at all with respect to the Sun. However, as recently as 1962, astronomers realized that Mercury's dark side was far too warm to be permanently turned away from the Sun; in other words, it was not a "dark side" at all. A few years later, radar echoes bounced off the planet from the Arecibo radio telescope in Puerto Rico produced the actual rotation period—58.6 days. So it is only 40 years since Mercury's rotation period was established: a testimony to how tricky it has been to record its surface features! So why should we tend to see the same surface features? Well, if an observer on Earth is observing Mercury at a favorable time (e.g., in the evening sky in a northern hemisphere spring), this favorable situation will next occur roughly three Mercury synodic (Mercury catch up time) periods later. Mercury catches Earth up every 116 days, and three synodic periods is just two weeks short of an Earth year. In addition, with the Earth in roughly the same part of its orbit, an Earth-based observer will be looking at a planet that has rotated roughly six times on its axis (6 × 58 days). The two "coincidences" here are that 3 × 116 is not far short of 365 days and 58 is half of 116.

Incidentally, there are various coincidences or, rather, orbital periodicities in the solar system, most of which are caused by gravitational/tidal forces. Eight Earth years equal approximately 13 Venusian years and two Uranian years equal approximately one Neptunian year. We shall see shortly that the moons of Jupiter are similarly linked.

Mercury Maps

We have already seen that Mercury, at its greatest elongation from the Sun, is a mere 7 or 8 arc-seconds in diameter. So what features, if any, are visible on the disc?. The British Astronomical Association, long associated with quality observations of all the planets, recommends using the albedo map produced at the IAU Planetary Data Center at Meudon, prepared by J.B. Murray, under the direction of A. Dollfus in 1971. Initially, Murray prepared a map based largely on high-resolution New Mexico State University Observatory photographs from 1965 to 1970. Visual observations from Lyot and Dollfus at Pic du Midi (1942–1966) were then added to several areas. The IAU coordinate system, recommended in 1970, was then added. Although much higher resolution Mariner 10 images exist from 1974, these are really too detailed (confusingly so) to show what Earth-based observers can record. The map in question, refined by David Graham of the British Astronomical Association (BAA), is reproduced in Figure 11.1. This was originally

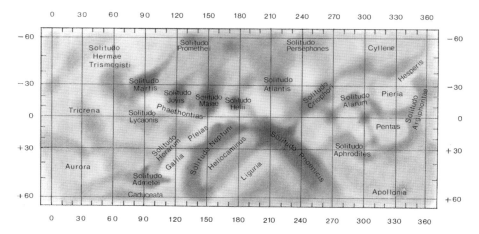

Figure 11.1. David Graham's map of Mercury: produced primarily for BAA observers, it is a refined version of the map produced by the IAU. Map: courtesy David Graham.

published in the February 1995 Journal of the British Astronomical Association. Of course, with webcams, the details glimpsed by the most eagle-eyed observers at Pic du Midi in the 1940s to 1960s can now be recorded with modest apertures, seeing permitting. More recently, the amateur astronomer Mario Frassati of Crescentino, Italy, produced an excellent map (Figure 11.2) for amateur observers based on his own visual observations of the planet, with a 203-mm Schmidt-Cassegrain. This too was reproduced in the Journal of the British Astronomical Association in June 2002. The drawings made to compile Mario's map were secured between January 1997 and May 2001. From a total of 78 high-quality drawings during that period, 54 were used to compile his map. Mario, his telescope, and his son are shown in Figure 11.3 and some of his excellent eyepiece sketches are shown in Figure 11.4.

Mercury is, of course, very close to the Sun and this produces its one single advantage when taking webcam images. The surface brightness is high. Mercury's albedo is only 6%, that is, it reflects 6% of the light landing on its surface. This is a similar albedo to that of our Moon. However, because Mercury orbits the Sun 2.6 times closer (on average) it has a surface brightness nearly 7 times greater than our Moon. This can be vital for clinching short exposures that freeze the atmospheric seeing. However, it should be remembered from our previous discussions about webcams that the actual exposure time indicated by the software can be false. In manual mode, an exposure at 5 or 10 frames per second will be 1/5th or 1/10th of a second, regardless of the indicated 1/25th exposure. Also, increasing the frame rate with a USB 1.1 webcam will lead to substantial image compression, which corrupts the data. We can use the brightness of Mercury (and the unfiltered Venus) to our advantage by reducing the exposure to, say 1/100th (while keeping the frame rate at 10 frames per second) and increasing the f-ratio to reduce interpixel noise to below 0.1 arc-seconds. Also, even with a narrow-band IR filter, the planet will still be bright.

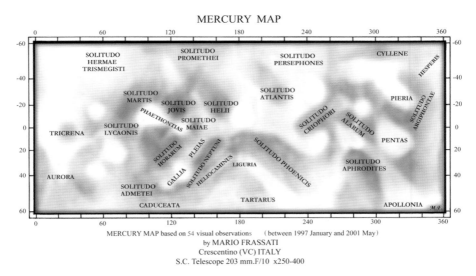

Figure 11.2. Mario Frassati's amazing map of Mercury, based purely on his own visual observations with a 203-mm Schmidt-Cassegrain.

Figure 11.3. Mercury mapper Mario Frassati and his son Lorenzo with Mario's 203-mm Schmidt-Cassegrain.

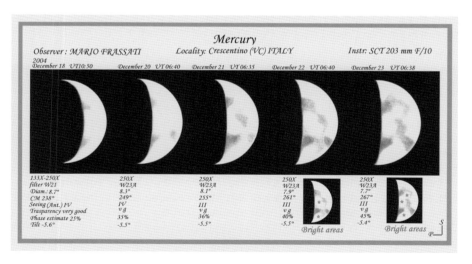

Figure 11.4. Sketches of Mercury made visually, at the eyepiece, by Mario Frassati with a 203-mm Schmidt-Cassegrain.

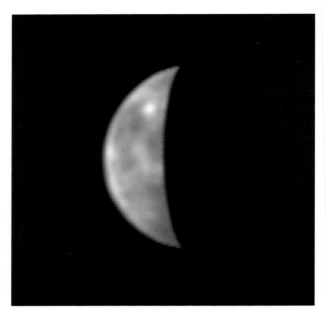

Figure 11.5. The best earth-based image of Mercury, obtained by the Boston University team of Jeffrey Baumgardner, Michael Mendillo, and Jody K. Wilson. An RS 170 surveillance camera was used at the f/16 focus of the 60-inch (152-cm) Mt, Wilson reflector. This produced an image scale of 0.09 arc-seconds per pixel. The best 60 images from 300,000 (1/60th second) exposures taken on August 29, 1999 were used to form this hi-res result. The planet was 7.7 arc-seconds in diameter at the time. Image: courtesy Jody Wilson.

The highest resolution image of Mercury obtained from Earth is shown in Figure 11.5. Although a 1.5-meter telescope was used, it is quite possible that amateur webcam users could rival this resolution given suitable seeing conditions.

Brilliant Venus

As far as the planet Venus is concerned, many of the factors that affect observations of Mercury are similar for Venus. It too orbits the Sun within the Earth's orbit and it is rarely high above the horizon after sunset. Like Mercury, it is biggest when merely a thin crescent (or when transiting the Sun) and it is best when elongated as far from the Sun as possible.

Venus is the planet most similar in size to the Earth. It has an equatorial diameter of 12,104 kilometers, i.e., 95% of the Earth's. Venus orbits the Sun in 224.7 days at a mean distance of 108.2 million kilometers and it overtakes the Earth on the inside track every 583.9 days (the synodic period). But if you thought Mercury's rotation period was weird, well, you could argue that the Venusian (or Cytherian) day was weirder. Venus rotates on its axis in 243 days. Yes, that's right. The day is longer than the year. Actually, this is of little interest to the webcam imager as we cannot see the Venusian surface. Venus has a dense atmosphere.

However, the upper clouds do have a retrograde (i.e., backward, like the planet's rotation) rotation period of about four days. Visually, Venus is both beautiful and featureless. Well, not entirely featureless, as observers with keen eyesight and experience can see some subtle atmospheric details in the atmosphere, especially if they have good sensitivity in the violet end of the spectrum and are using an appropriate filter. To me, Venus is at its most mesmerising when a crescent. It is nice and big (30 to 40 arc-seconds across when at a phase between 40% and 20%, respectively) and at a decent elongation from the Sun (45 to 35 degrees).

Despite the bland appearance of the dazzling Venusian cloud tops, there are a couple of relevant phenomena that are of particular interest to the amateur observer, although most of the interest is historical. The first feature is the phase anomaly, most obvious when Venus is supposed to be at half-phase, or dichotomy. Venus looks half when it is actually a bit fatter than that. Put another way, the illuminated part looks slimmer than it actually is. For evening apparitions, with the phase shrinking from gibbous to half to crescent, dichotomy appears to happen several days early. For morning apparitions dichotomy appears to happen a few days late. It should be stressed that, to the novice, the difference is very subtle, amounting to a phase difference of maybe 2%. With digital images there is no observer bias though. The effect is undoubtedly due to the Venusian atmosphere and it varies depending on what color filter is being employed.

The second phenomenon is far more controversial. It is called the "ashen light" and can only be seen when Venus is a crescent; usually a thin crescent. The ashen light is the term used to describe the alleged visibility of the nighttime side of Venus, or the dark side if you prefer. Of course, with the crescent Moon we see this phenomenon, called Earthshine, every month. The un-illuminated side of the Moon glows faintly due to light reflected from the Earth (especially the Earth's clouds) back onto the Moon. There is no mystery about the lunar Earthshine, the near-full Earth will shine as an object of magnitude −17 as seen from the Moon and will be as bright as 70 full moons would shine in our sky (because Earth is much larger and much more reflective). However, there is no way that the Earth can reflect a significant amount of light onto Venus. From Venus, a near-full Earth would only appear as an object of magnitude −6, nowhere near bright enough to illuminate the Venusian cloud tops. So what else could the ashen light be? Possibly the most popular idea is that it is simply an optical illusion, a contrast effect between the dazzling crescent and the nighttime sky. The brain just completes the rest of the circle. However, experienced observers are well aware of the tricks the eye and brain can play and they have used tiny curved occulting bars in the eyepiece to shield the dazzling crescent from view . . . and they have still reported the ashen light from time to time. But, despite this evidence, and rather disturbingly, even experienced observers have, from time to time, recorded a Venusian dark side that is actually *darker* than the background twilight sky. This, of course, is impossible to justify and must be due to an optical illusion. Only the ashen light observations in which a bright ashen light has been recorded by highly experienced observers, using an occulting bar, can be taken seriously. But if this small percentage of cases is genuine, what can be the cause? There appear to be only a few possible explanations, namely: violent thunderstorms in the Venusian atmosphere; intense Venusian aurorae, produced electrically; horizontal atmospheric scattering of light; observers who have exceptional eyesight in the near infrared;

Figure 11.6. The night side of Venus imaged by Christophe Pellier on May 19, 2004, with a 356-mm Celestron 14 SCT and ATiK 1HS webcam. A short exposure on the left captures the visual crescent and an infrared (1000nm) 105 × 8 second stack captures the dark side glowing. Image: C. Pellier.

or, an illusion. To me, this latter explanation is the only one that seems likely for the majority of cases. Until I see a CCD image of the Ashen light, I am inclined to regard it as an illusion, even though I actually know experienced and genuine observers who have seen this elusive phenomenon. CCD images of the dark side glowing in the infrared have been taken and these are discussed shortly.

Ultra Violet Venus Imaging

The cloud markings in the upper Venusian atmosphere are quite obvious features in the ultraviolet part of the spectrum (see Figure 11.7). Possibly the most popular filters for this type of work are photometric U-Band filters with a pass band centered on 365 nanometers. Such filters allow through light of wavelengths from 300 to 420 nanometers with a peak transmission at about 365. Photometric U-band filters have a great advantage in that they are coated to reject light from the near infrared that leaks through cheaper filters and to which CCD chips are especially sensitive. (Remember, when you remove the lens on a webcam you remove the standard UV and IR blocking filter built into the tiny lens.) A much cheaper alternative is a standard, photographic Wratten 47 filter. However, this filter will require an infra red blocking filter with it as most Wratten 47s will leak infrared through to the CCD. It is important not to get confused here with UV filters, IR filters, UV+IR blocking filters, and IR blocking filters! Blocking filters block, as their name suggests, and many amateurs already own UV+IR blocking filters so that their lensless webcams can be restricted to the same visual band as the human eye. This restriction reduces atmospheric dispersion effects and gives a more natural color. However, combining a Wratten 47 with a UV+IR blocker will be a disaster, as no light will get through! You need a photometric U-band filter *or* a stacked Wratten 47 and infrared blocking filter.

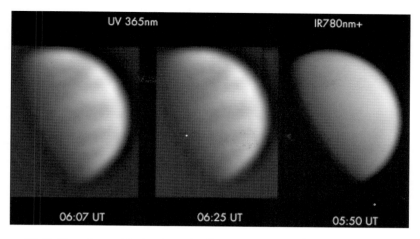

Figure 11.7. Venus imaged in the ultraviolet (first two images) and near-infrared (right-hand image) by Damian Peach on September 19, 2004, with a 280-mm Celestron 11.

CCDs are not very sensitive in the ultraviolet. In fact, their sensitivity drops virtually to zero as the wavelength drops to 300 nanometers. However, this is not too much of a problem as, at 365 nanometers (the peak of a U-band filter) CCD response is certainly low, but not zero and, crucially, Venus is a very bright planet. It is close to the Sun and has a high albedo. Venus' bright surface is a godsend when imaging in the UV. Nevertheless, using a color webcam, like a ToUcam Pro, will give far noisier results than using a sensitive, filtered, monochrome webcam like an ATiK 1HS.

If you do not have a UV filter, Venus is still a fascinating object to image in white light, where its brightness can allow very good signal-to-noise images with a short exposure time. Unlike planets like Mars, Jupiter, and even Saturn, the planet's rotation is not a limiting factor in obtaining thousands of frames to stack. The limiting factor is your stacking software and spare hard disk capacity.

Infrared Venus Imaging

Although Venusian upper cloud features are mainly ultraviolet features, some duskier markings are also visible in the near infrared. A filter with a bandpass of 700–1000 nanometers will reveal these markings. The U.K. imager Damian Peach has experimented with imaging in both UV and IR with Venus and then synthesizing a green component by adding the two results together. This can be used to make a false color RGB image consisting of infrared, green (IR+UV), and ultraviolet. The resulting image gives a picture that shows UV detail as deep blue and IR detail as red, a sort of highly exaggerated colour picture compressing Venus' subtle colors into the visual range. Professional astronomers used a similar technique when imaging Saturn's Moon Titan from the Cassini probe.

But what may interest amateur astronomers far more is the possibility of imaging the Venusian Ashen light in the near infrared. The reader of this book may well pose the question: "Surely someone must have tried this by now?" Indeed, they have. Both professionals and amateurs, but the attempts have been few and the visual Ashen light sightings are still hard to explain or discount. Probably the first observer to report the Ashen light was Johannes Riccioli in 1643. In the mid-1980s, Dr. David Allen and his colleagues at the Anglo-Australian Telescope showed that the Venusian night side did indeed glow strongly in the infrared. Unfortunately, it was far into the infrared, at wavelengths of 1.7 and 2.3 microns. However, do not give up hope just yet! The Galileo Near-InfraRed Mapping Spectrometer also imaged the Venusian night side (in 1991) at a wavelength of only 1.05 microns, far closer to the near infrared. The features discernible indicated that the detector was able to see the heat coming off features below the clouds, i.e., surface features. With the surface of Venus able to reach temperatures up to 460° Celsius in daytime, perhaps this is not surprising?

The amateur astronomer Christophe Pellier did achieve successful imaging of the Venusian night side in May 2004 (Figure 11.6). Using a Celestron 14 (355 mm aperture) at f/11 and an ATiK 1HS monochrome CCD he recorded an Ashen light type image with stacked 10-second exposures using a narrow-band 1000 nanometer filter. The night side was clearly visible, although there was only the vaguest suggestion of it using a 780–1100 nanometer filter on the same night. Maybe, you might think, this is the first evidence that the ashen light is real? However, the extreme red limit of the dark-adapted human eye is usually reckoned to be around 700 nanometers, so the idea that even superhuman observers could see to 1,000 nanometers still seems absurd: more research is needed.

In concluding this chapter, I would like to repeat, with some caution, that some amateur astronomers have imaged Venus in broad daylight. This is not difficult as, if you know exactly where to look, Venus can be spotted, in daylight, with the naked eye. Seeing in the daytime is usually much worse than at night, but the planet is at a much greater altitude, offsetting the poor daytime turbulence. However, for the beginner I would not recommend daytime imaging of Venus (or Mercury) at all. With the Sun in the sky and both planets quite close to it, this is a dangerous pursuit unless you have quite a few years of experience behind you.

CHAPTER TWELVE

Imaging Mars

Talk about the planet Mars and one immediately conjures up thoughts of science fiction: specifically, H.G. Wells' *War of the Worlds* and Orson Welles' radio dramatization of it (which scared a nation on October 30, 1938!). One also conjures up visions of the alleged canals on Mars, primarily "hyped" by Percival Lowell and the belief, even up to the mid-20th century, that intelligent life might exist on the red planet. The visual telescopic observers of the late 19th and early 20th centuries strained at the eyepiece to see any evidence of the Martian canals, but the damage done by the Earth's atmosphere meant that even with a lifetime of experience and good equipment most were still not sure if the canals were real or illusory. Only those few observers who had observed the planet under near-perfect conditions realized that canals simply did not exist. However, it took the fly-by of the spacecraft Mariner 4, in July 1965, to prove that Mars was a highly cratered world and not one able to support any intelligent life. The extent to which people believed that Martians might exist is hard to appreciate in the 21st century. However, the following anecdote, often quoted by Sir Patrick Moore, may make things clear. On December 17, 1900, a prize of 100,000 Francs (the Guzman prize) was offered to anyone who could establish contact with a being from another world. However, Mars was excluded from the award, because contacting a Martian was considered to be too easy!

Mars has an equatorial diameter of 6,794 kilometers (compared to 12,756 for the Earth). It orbits the Sun in just under 23 months (687 Earth days) and Earth overtakes, or laps Mars, on the inside track, in just under 26 months (780 Earth days). So, roughly three Earth months after each Martian year, we come closest to Mars. Mars orbits the Sun at an average distance of 228 million kilometers, roughly 50 percent further away than the Earth. However, the distance actually varies between 207 and 249 million kilometers. If Earth passes Mars when the red planet is closest to the Sun, it can come within 56 million kilometers. At such times (late

August/early September encounters) Mars can appear a whopping 25 arc-seconds across. Unfortunately for North American and European observers this takes place when the planet is low down in the southern sky. When Earth passes Mars at the red planet's furthest point from the Sun, it only spans 15 arc-seconds, quite a challenging object on which to resolve much detail.

Observing and Imaging Mars

Mars is by far the easiest planet on which the webcam user can record high-contrast detail. But for any imager a good map is essential, and a truly excellent one, aimed at the amateur observer, has been produced by Mario Frassati and Paolo Tanga. Their map is reproduced as Figure 12.1. It shows the planetary features exactly as they appear in a typical amateur telescope, although dust storms can greatly alter the planet's appearance.

Mars is a bright planet; not as bright as Mercury and Venus, but then those inner planets can never be seen as a fully illuminated disc when large, and they do not have the wealth of visible high-contrast surface features the red planet has; at least, not unless you own a spaceship! Mars has an albedo of 16%, which is a vast improvement on our own Moon's reflectivity (7%) and Mercury (6%). In addition, Mars' red color is a huge advantage to the webcam imager. I cannot stress this latter point enough. In typical poor seeing conditions a deep red filter will improve the seeing dramatically. A jittering, juddering image suddenly becomes a slow rippling one. This is of little use though if the object in question has most of its features in the blue end of the spectrum. Fortunately, Mars is at its best (highest contrast surface markings) when seen through a red filter, so a very nice LRGB image (luminance provided by a deep red filter) can be obtained.

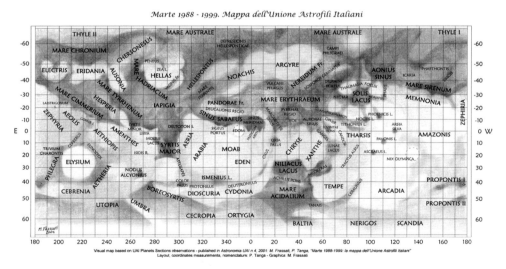

Figure 12.1. The excellent Mars map for telescopic observers "Mappa dell Unione Astrofili Italiani" produced by Mario Frassati and Paolo Tanga. By kind permission of Mario Frassati.

For observing visually, a deep red filter such as a Wratten 25A will greatly increase the contrast between dark and light Martian surface features and improve the atmospheric seeing, too. At the other end of the spectrum, a light blue filter, such as a Wratten 80A, will enhance clouds on the Martian limb. Mars is famous for its unpredictable dust storms that can encircle the whole planet at their most energetic. Under such circumstances, all surface markings can be rendered invisible.

Mars' Rotation

Mars has other webcam advantages, too. It has a relatively small disc compared to giant Jupiter and it rotates on its axis in 24 hours and 37 minutes (compare that with Jupiter's day of just less than 10 hours). This small size and slow rotation gives the webcam imager more time to capture images before resolvable detail in the center of the planet has smeared enough to be noticeable. How do you calculate the smearing rate? Well, let us pick a maximum allowable drift of the features on a planet's equator and meridian (i.e., the dead center of the disc) as 0.5 arcseconds. Admittedly, under perfect seeing conditions, higher resolutions might be achievable but, 0.5 is a reasonable value to start with. The formula we want is:

time window = drift limit / ((π × planet diameter) / rotation period)

The units for time window and rotation period must be the same, e.g., minutes. The units for drift limit and planet diameter must be the same, too, e.g., arcseconds.

Let us examine a few worked examples. For a 0.5 arc-second drift limit, the time window for a big, 25 arc-second Mars, rotating in 1477 minutes is: $0.5/((3.14 \times 25) / 1477) = 9.4$ minutes to collect the images. For a 0.5 arc-second drift and a smaller 15 arc-second Mars we get: $0.5/((3.14 \times 15) / 1477) = 15.7$ minutes to collect the images.

Contrast this with Jupiter at opposition with, say, a diameter of 45 arc-seconds and a rotation period of 590 minutes: $0.5/((3.14 \times 45)/590) =$ a scant 2.1 minutes to collect the images.

With so much time available, the webcam imager can obtain very sharp results just by using a simple ToUcam webcam and a deep red filter. Also, Mars is so bright that a monochrome webcam and a complete filter set are unnecessary. I obtained some nice images of Mars in 2003 by constructing an LRGB image with just one normal color webcam AVI file and an extra red-filtered AVI file with the same webcam. There was enough time to take the deep red filtered webcam image and then remove the webcam/Barlow unit and replace the red filter with a UV-IR reject filter. I ended up with a blurry color image and a high-contrast, deep red filtered image. By splitting the color image into its RGB components and using the red filtered image as the "L" or luminance monochrome component, the transformation was dramatic. All by using one filter and exploiting Mars' slow rotation and small size.

We can see, compared to imaging Jupiter, imaging Mars is an almost leisurely pursuit! However, the slow rotation of Mars can be frustrating, too.

If there is one planet that needs a worldwide network of amateur observers, it is Mars. As we have seen, the Martian day is only 37 minutes longer than our own, so astronomers around the world can only view one hemisphere at a time. By the time the other Martian side has rotated into view the planet has set and it is day-time. To see an entire Martian rotation from one site you have to observe the planet for five weeks! However, this does mean that a few cloudy nights will not be a problem: you will not miss much. To the novice, observing Mars through an astronomical telescope, things can be quite confusing. In an inverting telescope, with south at the top, Mars will slowly rotate from right to left, with the morning terminator on the right and the evening terminator on the left. Look at Mars a day later, at the same time and you will see 10 degrees further as features "emerge" from the left-hand evening terminator, as if the planet was rotating backwards! It is an illusion of course: you are simply seeing features on Mars slightly earlier in the Martian day, but it can be very confusing.

The Martian Features

Mars has a number of major dark markings on its surface, which, with a bit of patience, will yield to scrutiny from even a novice observer. Six "faces" of Mars, imaged by the author, are shown in Figure 12.2. For the beginner, the huge V-shaped Syrtis Major is the most obvious feature but other dark markings like Sinus Meridiani and Solis Lacus are also distinctive. With practice, your eye and brain will become trained to see more detail, detail that emerges as if by magic when image-processing routines are employed on stacked webcam images.

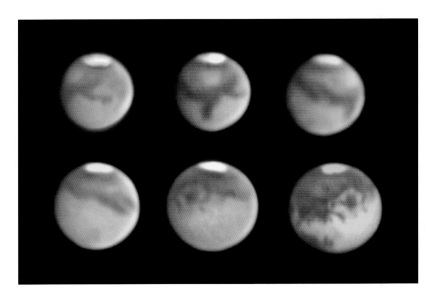

Figure 12.2. Almost a whole rotation of Mars is captured in this sequence of webcam images from top left to bottom right shot by the author over a four-week period using a 30-cm LX200 at f/22 in July and August 2003. South is at the top and roughly 60 degrees more of the planet emerges from the left-hand evening terminator in each frame, giving a night to night illusion of the planet rotating backwards! The major features in the middle of each frame are the Sinus Meridiani; the Syrtis Major; Mare Cimmerium; the Amazonis Desert; Solis Lacus to the left of center; Solis Lacus to the right of center.

I think the best way to describe the Martian features to a beginner is to imagine they are drifting into view escaping from the evening terminator with south at the top, in other words, in the opposite direction to the planet's rotation and as they would appear over a five-week period, to the observer. It is worth noting that a beginner will need some time to identify the major features visually as atmospheric stability (in the *Earth's* atmosphere!) is essential for resolving fine details. Mars is often a very small body and, perhaps surprisingly, can vary considerably in appearance. Remember, Mars has an axial tilt of almost 24° (like the Earth), so sometimes we get a better view of the southern hemisphere, and sometimes the northern hemisphere is favorably presented; sometimes both hemispheres are equally favored. Also, depending on whether it is Martian spring, summer, autumn, or winter in the hemisphere we are looking at, the relevant polar cap may be large, shrinking rapidly, or small. The axial tilt is crucial to how the major features at high latitudes appear; especially the polar caps. Add to this a disc that may only be a few arc-seconds in diameter, or may span 25-arc-seconds, and the extra factor of potentially considerable dust storm activity and you can see how Mars looks very different every year you observe it. On top of this is the problem I have already mentioned, i.e., that you may not see features on the opposite side of the planet for over a month, simply because Mars rotates in 24 hours and 37 minutes. While an observer in Florida is studying, say, the Syrtis Major, his colleague in Hong Kong may well, 12 hours later, be studying the Solis Lacus, but neither will be able to see the other's subject well for several weeks.

If we start first with the Syrtis Major, well, I have to admit a psychiatric problem here. To me, the Syrtis Major and adjoining features always look like a small bat clinging to an orange! Above (to the south) of this huge dark V is a noticeably lighter oval feature called Hellas. Many dust storms seem to originate first in this area. As the days tick by, observing at the same time of night, you will see more features emerging from the left-hand evening terminator. After a week or less you will notice a rarity of dark albedo features emerging, except for a dark stripe moving upward away from the Srytis Major region. This is the Mare Tyrrhenum/Mare Cimmerium region and marks the start of what, as a Mars novice in the 1980s, I used to call "the boring side of Mars"! By that I meant the hemisphere with a rarity of dark albedo features, except in the high southern latitudes. After a week or more of tolerating this "boring side," the beginner will start to see a fascinating feature emerge from the evening terminator. This is the Solis Lacus or "Lake of the Sun." At first it just looks like a dark smudge on the limb, but when on the meridian it resembles a dark eye or the hub of a wheel, with a hint of spokes radiating out from its center. Beyond and bordering the Solis Lacus' lighter surrounds is Aurorae Sinus and then, way down in the north a rare dark northern feature, the Mare Acidalium swings into view. Only a few days later you will see another distinctive marking emerge. It is called the Sinus Meridiani, and to its left, a long dark line is seen, the Sinus Sabaeus. To me (and remember my problem with a bat clinging to an orange!) the Sabaeus-Meridiani feature looks like the arm of a bear with a claw (Meridiani) at the end. Meridiani is so-called because it is at the point designated as zero degrees longitude on the Martian globe. Finally, a few days after the whole of Sinus Sibaeus emerges we are back to the giant dark V of the Syrtis Major, emerging from the evening terminator; and we have now been observing Mars for over a month.

Mars' expanding and shrinking polar caps are unique features for the Earth-based observer to study, but they can also be a real problem for the webcam imager as they are simply so much brighter than anything else on the planet (see Figures 12.3 and 12.4). The skilled visual observer still has a slight advantage here as the eye-brain combination can tolerate a much larger brightness range than an 8-bit webcam frame. Sometimes the only solution to the polar cap brightness is to image the planet at two different exposures simply to reveal any detail within the caps themselves. Features within the caps are rare, but they do occur, at the cap boundaries, when local spring causes them to shrink. Typically, with the southern cap, two rifts have appeared (Rima Australis and Rima Angusta) and then a detached part of the cap is left. (Schiaparelli called this Novissima Thyle). At a later stage this too breaks up into the so-called "Mountains of Mitchel," first seen by O.M. Mitchel at the Cincinnati Observatory in 1845. Similar changes can be recorded in the northern cap, too. With the onset of autumn, an overlying polar haze can confuse the situation. But, regardless of all this, the polar caps are dazzlingly bright and a combination of short exposures, filters and careful use of brightness, contrast and gamma controls are needed to produce images that show the Martian features and the polar cap edges well.

Occasionally some concern has been voiced that the Martian colors, obtained by using a color webcam like the ToUcam, are not exactly true to life and well below the purity that could be obtained using a true RGB filter set. The filters built into

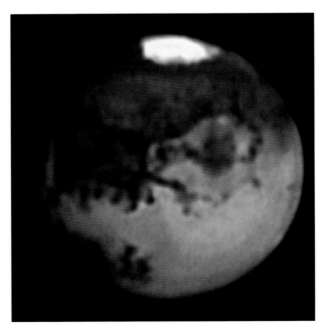

Figure 12.3. Mars imaged on August 22, 2003, by Damian Peach, at high altitude, from La Palma in the Canary Islands. A 25-cm Schmidt-Cassegrain working at f/40 was used, as well as a ToUcam Pro webcam. The major dark "eye" above and to the right is the Solis Lacus.

Figure 12.4. Mars imaged one day later than in Figure 12.3, and 30 minutes earlier. The Solis Lacus is on the right hand of the disc, Aurorae Sinus near the center, and Sinus Meridiani on the left hand of the disc.

the ToUcam pixels (one blue, two green, one red in each 2 × 2 pixel cluster) cover a wider bandwidth than narrowband scientific filters, so they can grab more signal. Damian Peach and Singapore imager Tan Wei Leong studied this issue some time ago and concluded that the use of a UV-IR rejection filter combined, crucially with a magenta filter (effectively a green subtraction filter) would lead to better filtering and the final ToUcam image could then be cleanly separated into red and blue channels with green being created from red + blue to synthesize an aesthetically pleasing color picture, too. However, with the advent of low-cost monochrome webcams and filter sets, this technique has rarely been used. When one takes into account atmospheric dispersion, individual telescope characteristics and individual observer's picture tweaking methods, one can conclude that the only truly scientific images are monochrome ones taken through precision filters before they are combined to make a pleasing color image. However, another approach in imaging Mars, pioneered by Antonio Cidadao in Portugal, is to image the planet in IR and UV and synthesize the green channel completely, by combining the IR and UV images. See Figure 12.5 for the result that captures the surface details and high-altitude limb haze. A more standard approach, by Don Parker, is shown in Figure 12.6.

Mars is often neglected by amateur astronomers when the disc shrinks to less than 10 arc-seconds in diameter. However, good views can still be obtained when the planet is high up and seeing is good. Take a look at the stunning image by Damian Peach, in Figure 12.7, for proof.

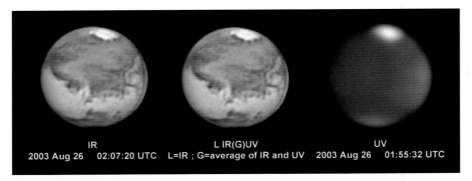

Figure 12.5. Antonio Cidadao from Portugal pioneered the technique of simulating the green channel in planetary imaging. Here the technique is used on Mars. The infra-red image captures the Martian surface markings well and the ultraviolet image captures the limb haze. By combining the IR and UV images to give a synthetic green, a clean color image of Mars showing everything of interest is captured, and only two filters are needed. A 250-mm SCT with Stellar Products adaptive optics device was used, alongside a Finger-Lakes Instruments CM7-1E CCD. Image: A. Cidadao.

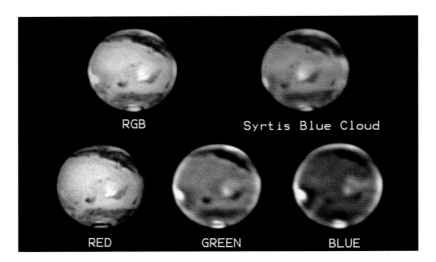

Figure 12.6. Mars imaged by Donald Parker of Coral Gables, Florida, on April 24th 1999, with his 40-cm f/6 Newtonian and a Lynxx PC CCD camera through red, green, and blue filters. Note the different appearance of the planet in each color. The images show a "cold front" coming off the North Polar Cap, a brilliant orographic cloud over Olympus Mons on the evening (left) limb, and clouds over the Elysium volcanic shield near the center of the disc. Also note the famous Blue Syrtis Cloud on the morning limb, in a slightly later enhanced color shot in the top right image.

DECEMBER 18th, 2003
CML=251, Ls=318
17:11 UT

Dia=9.40"
D. Peach

Figure 12.7. This amazingly hi-res image of Mars was obtained from the U.K. by Damian Peach, on December 18, 2003, in a twilight sky. A Celestron 11 telescope and ATiK 1 HS webcam were used. The resolution is slightly better than one would expect for an instrument of this aperture: note the arc-second scale down the side and on the planet! Image: Damian Peach.

The Martian Moons

A real challenge for the amateur observer, even at a favorable opposition, is spotting the Martian moons Phobos and Deimos. In a large amateur telescope these 11th and 12th magnitude moons would not normally be a challenge, but they never stray far from the brilliant Martian disc and their rapid orbits mean they move considerably even in one night. Phobos is the larger Moon at over 20 kilometers across but never strays more than a Martian diameter from the planet, making it the hardest to see. Deimos, at around 12 kilometers across can stray three Mars diameters from the planet, making the fainter Moon easier to see. The best plan for the visual observer is to use an eyepiece with a thin "occulting bar" at the eyepiece focus. Placing Mars behind the bar reduces most of the dazzling light enabling both moons to be seen. However, the exact position of the moons needs to be checked each night with a planetarium software package such as Guide 8.0. Phobos and Deimos orbit the red planet in 8 and 29 hours, respectively.

Of course, with a webcam, using an occulting bar is not essential, because a webcam chip, unlike the eye, cannot be dazzled (although Mars will certainly end up being saturated and burnt out). Because Phobos and Deimos orbit so close to Mars, fitting the planet and moons in the same image is not hard, and, to make a nice composite, you can easily cut and paste a short exposure Mars image on top of the overexposed blur in the main image. It might be thought that a webcam could not reach objects as faint as 11th or 12th magnitude with short exposures, but in fact they can. (An alternative solution with ATiK or modified webcams is to take a long exposure beyond the normal 1/5th second limit of a commercial webcam.) Even a humble ToUcam Pro can just record stars of magnitude 11 or so in 0.1 second exposures at f/10 with a 25-cm aperture. With a monochrome webcam and 0.2 second exposures stars as faint as magnitude 12 or 13 can be recorded on a stack of hundreds of frames.

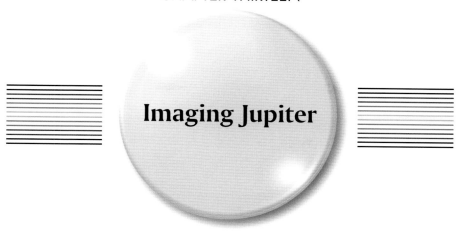

CHAPTER THIRTEEN

Imaging Jupiter

Of all the planets in the solar system, Jupiter is the most rewarding to study because changes can be glimpsed from night to night as the huge weather systems move with respect to one another. I never tire of seeing the Great Red Spot appear over the limb or the satellites and their shadows crossing the giant planet. Jupiter is the fastest rotating planet in the solar system and even a novice observer will spot the features moving across the disk in a half-hour session. Jupiter's huge invisible shadow, trailing behind the planet, creates a fascination of its own as the four Galilean moons drift in and out of the shadow and sometimes even eclipse and occult each other. On nights of perfect stability, small markings can just be glimpsed on these moons.

Jupiter is truly massive, with an equatorial diameter of 142,880 kilometers, that is, just over 11 times the Earth's diameter. Even though the planet is mainly gas and liquid, with (probably) a relatively small rocky core, it still has a mass of some 318 Earths! Jupiter's equatorial regions (System I) rotate in 9 hours, 50 minutes, and 30 seconds, with the rest of the visible regions (System II) taking 5 minutes, 11 seconds longer. A third rotation system (System III), based on radio signals from the planet, gives the rotation period as 9 hours, 55 minutes, and 29 seconds. With there being no visible solid surface, the system longitudes transiting the meridian at any given time need to be calculated without reference to solid features. Thus, the longitudes of certain features (such as the Great Red Spot) tend to drift slowly with respect to the system they are in. Planetarium packages like Project Pluto's Guide 8 or handbooks like that of the BAA can be used to check on the system longitudes at a given time. Jupiter orbits the Sun at a mean distance of 778 million miles. At its closest it can appear as large as 50.1 arc-seconds in diameter, although an opposition size of nearer 45 arc-seconds is more typical. The giant planet orbits the Sun every 11.86 years and so spends half that time above the ecliptic (good for northern hemisphere observers) and half below (good for southern hemisphere observers). As always, observers on the equator have no reason to ever complain! Jupiter reaches opposition 32 days later each year. The belts, zones, and jet stream currents on the giant planet are highly complex, but hopefully Figure 13.1 will make things slightly clearer than mud! Figures 13.2 through to 13.6 show some high resolution webcam images of Jupiter.

The Galilean Moons

Jupiter's four giant moons (Io, Europa, Ganymede, and Callisto) are useful targets for the webcam imager and for a variety of reasons. Jupiter is a low-contrast gaseous body with no high-contrast features like the Martian Syrtis Major, and no sharp-edged features like Saturn's rings. So what can you focus on? The planet's features and limb are just too fuzzy most of the time, unless you are using an infrared filter, but Io makes an excellent target. It is always relatively close to the planet (a quick search east or west will easily find it unless it is in front of or behind Jupiter) and is bright enough for a webcam on high gain to pick up. Yes, I know it is not a point source like a star, but, nevertheless, focusing back and forth on Io until you are happy it is as small a dot as you can get it is a good plan for the novice imager. The other Galilean moons can also be used, but they are usually much further away from the disc. In order of distance from Jupiter, the satellites Io, Europa, Ganymede, and Callisto orbit at a maximum angular distance of 2.3,

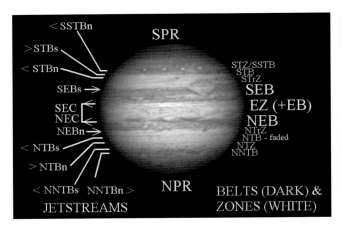

Figure 13.1. Jupiter's complex system of belts, zones, and currents, with the Great Red Spot to the right of the central meridian. The belts are dark features and the zones are bright features. The arrowheads indicate the direction of the jetstream currents on the planet which tend to lie along zone/belt boundaries. The Great Red Spot vortex rotates anti-clockwise. Jupiter Image: D. Peach. Legend from top, clockwise: SPR = South Polar Region; STZ/SSTB = South Temperate Zone/South South Temperate Belt; STB = South Temperate Belt; STrZ = South Tropical Zone; SEB = South Equatorial Belt; EZ + EB = Equatorial Zone + Belt; NEB = North Equatorial Belt; NTrZ = North Tropical Zone; NTB = North Temperate Belt; NTZ = North Temperate Zone; NNTB = North North Temperate Belt; NPR = North Polar Region. On the left hand side the terminology is the same, with s and n indicating North and South belt edge jetstream currents. NEC/SEC = North/South Equatorial Current.

Figure 13.2. A highly detailed webcam image of Jupiter taken by the veteran Japanese observer Isao Miyazaki on February 28, 2004, using a ToUcam Pro and his 40-cm f/6 Newtonian. The long-lived oval "BA" is seen in the top left. Image: I. Miyazaki.

Figure 13.3. Another webcam image of Jupiter by Isao Miyazaki, taken on March 11, 2004. Oval BA is in the top left quadrant and the Great Red Spot near the top left limb. The shadow of the Moon Callisto is about to leave the planet's disc. Image: I. Miyazaki.

Figure 13.4. A Jupiter image by the author taken on April 11, 2004, using a 30-m LX200 at f/22 and a ToUcam Pro webcam; 366 webcam frames were stacked. Image: M. Mobberley.

Figure 13.5. An excellent ToUcam Pro webcam image stack of Jupiter taken by Eric Ng from Hong Kong using a 250-mm f/6 Newtonian (at f/34.5) with William Royce Optics and a Vixen Atlux mounting. March 17, 2003. Image: Eric Ng.

Figure 13.6. One of Damian's finest images of Jupiter, taken on March 4, 2003 (23:29 UT), with a Celestron 11 at f/30 and a ToUcam Pro webcam. Io and its shadow can be seen on the planet. Image: Damian Peach.

3.7, 5.8, and 10.3 arc minutes respectively from the center of the planet's disc at a typical opposition (with Jupiter being roughly 45 arc-seconds in diameter). The time taken for each Moon to complete an orbit is 1.8, 3.5, 7.2, and 16.8 days, respectively. In the same order, the moons' diameters are 3,650, 3,130, 5,268 and, 4,806 kilometers, thus they have an angular diameter of 1.2, 1.0, 1.7, and 1.6 arc-seconds at an average opposition of Jupiter with the giant planet 4.2 A.U. from Earth. This tells us straight away that it should be possible, in good conditions, to resolve details on the Jovian Moons and, indeed, this is possible, as shown in Figure 13.7.

Observing and imaging the Galilean moons, as they transit the planet's disc, can be a challenging feat. Apart from when a Moon is passing over the darker limb regions, most of them will have a similar albedo to Jupiter's bright zones and, unless seeing is perfect, the Moon will become lost in the Jovian background. The one exception to this is Callisto, which is by far the darkest Moon with an albedo of 20%. (Io, Europa and Ganymede have albedos of 61, 64, and 42% compared to Jupiter's average albedo of 43%.) The first time you see Callisto transiting the Jovian disc you will be convinced it is a shadow of one of the moons! However, you will not see it cross the disc very often; with an orbital period of 16.8 days, it only spends a few hours per fortnight crossing the disk, if it crosses it at all (being 1.9 million kilometers from Jupiter, the orbital tilt often causes it to miss the planet entirely). Savour those moments when you see dark Callisto crossing the Jovian disc!

Another question which all this brings up is as follows: if we set a maximum limit of two minutes of time for collecting our Jupiter images (equivalent to almost

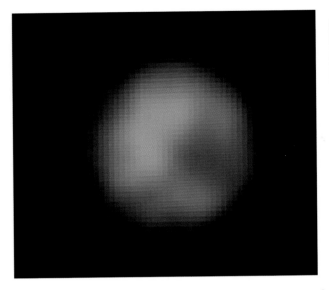

Figure 13.7. An extraordinary high-resolution image of Jupiter's largest Moon Ganymede, taken with a 280-mm Celestron 11 working at f/30, from Tenerife, with a ToUcam Pro webcam. The Moon has an angular diameter of only 1.7 arc-seconds! Image: Damian Peach.

0.5 arc-seconds of rotation drift on the middle of the disc) is this short enough to prevent the moons (or their shadows) elongating during the exposure run?

Anyone who enjoys observing Jovian satellite and shadow transits will have noticed that when Io, Europa, or Ganymede are transiting the central part of the giant planet's disc they appear almost to be part of the planet: atmosphere and satellite (on the face of it) appear to move at a similar rate! The most frequent transiting body is Io as it orbits in 1.77 days at 422,000 kilometers from the planet. The corresponding figures for Europa, Ganymede and Callisto are 3.55/671,000, 7.16/1,070,000, and 16.8/1,880,000. At mean opposition distances from Earth these four orbital speeds correspond to motions of 20, 16, 12.7, and 9.5 arc-seconds per hour, compared to the Jovian center's speed of 16 arc-seconds per hour. So, in a two minute imaging span at opposition, fast mover Io will drift two-thirds of an arc-second, with respect to the Jovian limb, but much more slowly with respect to the fast-moving central Jovian disc features. Essentially, in a two minute period the motion of Io (with respect to the Jovian limb) is roughly half Io's own diameter. This is irrelevant on all but the best resolution nights unless longer imaging spans are attempted to get more images to stack. Of course, if any of the four Galilean satellites are travelling over regions very near to the planet's limb, their relative motion (with respect to surface features) will appear faster due to foreshortening or the planet's smaller diameter at the poles. However, none of the big moons zip across the Jovian disc in the typical duration of an imaging window. Having said this, I have seen one very fine image by Damian Peach, using filters, where a double Io was visible because four minutes elapsed between the start of the first filter image (infrared) and the end of the second (blue). So Io can be seen to move in longer imaging runs, when seeing is good. Of course, to resolve detail on the satellites, Registax needs to be told to use the satellite itself as the reference stacking target.

Every six years or so the modest tilt of Jupiter's polar axis is aligned at right angles to the sun-Jupiter line (just as our Earth's polar axis is aligned at the spring and autumnal equinoxes). This results in all of the Jovian moons orbits appearing to be lined up, like looking at their orbital planes as if we were looking at the edge of a sheet of paper. This creates some interesting "photo opportunities" as the Jovian moons can mutually occult and eclipse one another. Under perfect seeing conditions it is possible to resolve details finely enough to show the shadow of one Moon passing over the other, or one Moon's limb cutting in front of another, as shown in Figure 13.8.

Is it possible to see two, three, or even four Jovian moons (and, more easily, their shadows) crossing the giant planet's disc simultaneously? Well, seeing two shadows crossing the disc is quite a common event if you have clear skies every night. However, in practice, how many people do? In reality, you might only observe such an event a few times a year if you are a keen observer, but with mainly cloudy skies. However, three shadows is practically a once-in-a-lifetime event. The student of Jupiter soon realizes that there are certain patterns in the revolution periods of the Jovian moons, just as there are in the Sun's own family of planets. Two revolutions of Io around Jupiter equal one revolution of Europa, and two revolutions of Europa equals one revolution of Ganymede. However, the pattern does not repeat for Ganymede and Callisto. The latter's revolution period is actually 2.3 times the former. These relationships mean that certain events involving the Jovian satellites repeat themselves every 3.6 or 7.2 days (the revolution periods of Europa and Ganymede). Even the novice observer notices the 7.2 day period pretty quickly. "Surely this same Europa-Ganymede event happened on this night last week?" you may find yourself asking. Yes, it did, but this week it is about three or four hours later. The 3.6 day Io-Europa repeat is harder to spot because the 0.6 day fraction tends to push a repeat performance into daylight

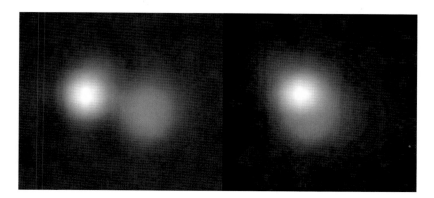

Figure 13.8. Another amazing image by Damian. In this webcam shot, taken on December 24th 2002, the Jovian Moon Callisto is being occulted by Io, i.e., Io is passing in front of Callisto. Io spans 1.2 arc-seconds and Callisto 1.6 arc-seconds. Left image at 03:06 UT; right image at 03:10 UT. Celestron 11 at f/30 and ToUcam Pro webcam. Image: Damian Peach.

hours, plus, as many observers only find time to observe at weekends a repeat a week apart is much more likely to be noticed!

As I have already mentioned, seeing three satellite shadows cross the Jovian disc is very rare, and indeed, four simultaneous transits are impossible: Io, Europa, and Ganymede can never be simultaneously involved in these events and so triple events have to involve Callisto and two of the inner three moons. The Belgian mathematician and expert in spherical and mathematical astronomy, Jean Meeus, has painstakingly calculated all triple shadow events from 1900 to 2100 A.D. in his remarkable book *Mathematical Astronomy Morsels* (plus, there are sequels: *More Mathematical Astronomy Morsels* and *Mathematical Astronomy Morsels 3*). I would strongly recommend buying these fascinating books. Unfortunately, the next three events are some way off, specifically on October 12, 2013; June 3, 2014; and January 24, 2015. After that, you have to wait until March 20, 2032! As well as the situation of three satellite shadows crossing the Jovian disc, three satellites themselves can also appear on the disc at the same time. I have seen one such event: the triple transit of January 17/18, 2003. At its peak, five objects were on the Jovian disk simultaneously (see Figure 13.9 for an image by Damian Peach from that night). These five objects were the shadows of Io and Europa, plus Io and Europa themselves and dark Callisto, too. Three moons and two shadows! Actually, as I was observing in poor seeing, it looked like a triple shadow transit, as Callisto is so dark and bright Io and Europa were invisible to me. Capturing such rare events is a great challenge from a predominantly cloudy country and you will savour your memories and pictures of such occasions.

Figure 13.9. A very rare Jupiter image. Taken January 17, 2003 (23:39 UT), with a 280-mm aperture Celestron 11 SCT from Tenerife, it shows, from left to right on the planet, dark Callisto, Io's shadow, Io, Europa's shadow and, just off the limb, Europa itself. Note how the dark Moon Callisto looks like a shadow. Image: Damian Peach.

Jovian Weather Systems

A detailed description of the weather systems of the Giant Planet are beyond the scope of this book. The outstanding book *The Giant Planet Jupiter* by John Rogers (Jupiter Section Director of the British Astronomical Association) is the definitive observers guide and acquiring a copy is a *must* for the keen planetary observer. Unlike any other planet in the solar system, Jupiter always has a wealth of changing detail that is visible in modest telescopes. The detail on Saturn is very subtle (except when a large spot comes along) and Martian weather is usually a subtle business, too, except for major dust storms and the shrinking and growing of the polar caps.

The easiest Jovian feature to observe (and easily the most famous) is the Great Red Spot or GRS. This rotating storm, bigger than the Earth, was probably first spotted in 1665 by Cassini in Italy although there is some speculation as to whether Robert Hooke of England observed it a year earlier. However, both of those features may not have been the GRS at all: it is just impossible to be sure. The earliest undisputed observation of this remarkable feature was in 1831 by S.H. Schwabe. The Great Red Spot is in the SEB or South Equatorial Belt and thus has a longitude in System II. However, the 9 hours, 55 minutes, and 41 seconds is only a mean figure for all the nonequatorial features and so the longitude of the GRS does drift with time. At the time of writing (2005) it is close to 100° but slowly increasing. Indeed, over the last 130 years it has moved backwards and forwards in System II, going around the globe five times! From about 1870 to the late 1930s it drifted backwards in System II almost five times. Then, in the last 60 years it has crept forwards, but at a much slower pace. From 1878 to 1882 the Great Red Spot became very prominent and brick red in color. At one point it became 40,000 kilometers wide, literally double its current width.

Monitoring the drift rate and evolution of Jupiter's Great Red Spots and the numerous other smaller spots is a major feature of Jupiter observing. The method that used to be employed to determine the longitude of a spot was to time it crossing the Jovian meridian. However, with webcam images being of such a high quality, it is just as accurate to pinpoint a feature's position with a cursor and measure it straight off, whether or not it is transiting the meridian. Perhaps the most advanced system for determining system longitudes on Jupiter is the JUPOS software developed by Hans Joerg Mettig for the British Astronomical Association.

To understand the weather systems of Jupiter it is important to start by learning that the planet can be divided up into a series of dark belts and light zones. The complete novice may only be able to identify the dark SEB (South Equatorial Belt) and NEB (North Equatorial Belt) and the dark polar regions (SPR and NPR). In addition the light Equatorial Zone (EZ) and the lighter regions between the polar regions and the NEB and SEB can be spotted even through a small telescope. Needless to say, things are a lot more complicated than this and the planet can be divided into 20 belts and zones when high-resolution images are analyzed. (see Figure 13.1)

But this is not an end to matters. The advanced observer will want to understand not only how to recognize the belts and zones but also how to expect features to drift within them. When looking at a satellite image of Earth you can spot the weather fronts but the prevailing air and weather currents are not immediately obvious. Jupiter has a system of slow currents associated with the most obvious features but it also has numerous prograding (with the planet's rotation) and retrograding (against the planet's rotation) jet streams at the boundaries of these obvious features. When looking at reports of Jupiter observers observations you may be confused by two issues. Firstly, planetary images are nearly always shown with south at the top (because astronomical telescopes produce an upside-down image in the eyepiece). Secondly, the terms p and f are frequently used. These stand for preceding and following. In other words, one object precedes another if it rotates into the observer's view first, as the planet rotates, and vice versa.

When we look at Jupiter's creamy zones, brownish belts, and other features, what, precisely, are we looking at? The Jovian atmosphere consists mainly of hydrogen (90%) and helium makes up almost all of the rest. So where do the colors we see come from? This is not a question that has a simple answer, indeed, in John Rogers' definitive book *The Giant Planet Jupiter*, 20 pages are devoted to discussing the vertical structure of the atmosphere, the colors seen by the observer and the "cloud" features. Even after the Galileo spacecraft mission, it is not possible to easily summarize what you are looking at when you see the Jovian disc on your PC screen, or through the eyepiece. Perhaps surprisingly, spectroscopy does not automatically provide a solution to the question of what the eyes see. The situation is highly complicated. However, some simplification is possible. The colors seen on Jupiter arise from the miniscule amounts of elements that are not hydrogen or helium.

The white clouds that make up the bright zones on the giant planet seem to consist of ammonia ice crystals floating in gaseous hydrogen. (Gases heated by the considerable internal heat of the giant planet, rise into the upper hydrogen atmosphere and cool.) These bright zones appear to be higher and colder than the dark belts.

The red/brown, gray/brown, or simply brown colors that characterize the SEB and NEB may well be made from ammonium hydrosulphide polymers.

The Great Red Spot is often the most obvious red or orange feature on the Jovian surface and the color may be caused by the condensation of phosphorus. Amateur webcam images are now of sufficiently high resolution that they routinely show the rotation of the spot itself. The GRS vortex, featuring wind speeds up to 360 kilometers per hour, rotates counter clockwise. On the outer edge one rotation is completed every 12 days, and on the inner edge one rotation is completed every 9 days. This vortex appears to be a high-pressure area that has risen by 8 kilometers above the surrounding clouds, because of convection from below.

As well as the major Jovian belts and zones, Jupiter has smaller, more subtle features whose terminology can be confusing to the beginner. The five most likely confusing or ambiguous terms encountered are festoons, barges, ovals, spots, and portholes.

A festoon is a dark, usually bluish, feature that projects into the white equatorial zone from the NEB south edge, and trails backwards against the direction of the planet's rotation. Variations on this theme exist, such as arched festoons, plumes, or simply projections.

A barge is a term first used by the Jupiter observer Captain Ainslie in 1917/18. It refers to dark oblong features with a distinct "chocolate red" or "clotted blood" coloration. To Ainslie, they looked like canal barges stranded by a falling tide on the NEB north edge. White ovals is a term generally used to describe long-lived features on the planet, whereas the term *white spot* generally applies to smaller, short-lived features. After the long-lived Great Red Spot, the three white ovals called BC, DE, and FA were probably the longest-lived features (but much smaller than the GRS) that have been studied since the invention of the telescope and the era of serious Jovian study. These white ovals dominated the activity in the South Temperate Region of Jupiter from 1940 until the end of the 20th century. The term *portholes* is often used to describe anticyclonic white ovals in a dark belt such as the NEB.

Major Jovian Upheavals and Events

From time to time, major developments take place in the Jovian atmosphere that can cause the planet's appearance to change dramatically over a period of years. A quick look at photographs of the planet taken over the last 100 years reveal how much things can change from decade to decade and often even more quickly. The Great Red Spot is about half the length now that it was at the dawn of astronomical photography. Major events that will attract the keen webcam user's interest will include the merging of spots and barges, and the interaction of spots with the Great Red Spot's whirling vortex. But much bigger planet encircling events can take place, and spotting the first signs of these developing is highly important (not to mention prestigious for the amateur). The Great Red Spot straddles the southern part of the dark (gray, brown, or reddish) SEB and the lighter South Tropical Zone (or StropZ as it is sometimes called). However, on a regular basis (every five years or so is typical) the SEB fades over a period of months until the southern component almost disappears. Usually, this fade lasts for one to three years before there is an SEB revival. While the SEB is faint, the GRS tends to become darker and redder. The revival of the SEB always starts with a dark spot or a dark streak lying across the SEB. Sometimes a bright spot is seen preceding the dark spot/streak, in other words, preceding in time and longitude. Dark revived belt coloration then spreads out to restore the SEB to its former state.

At similar latitudes, the South Tropical Zone, in which the southern half of the GRS resides, can suffer its own events. These are known as South Tropical Disturbances. There was a *Great* South Tropical Disturbance from 1901 to 1939, but six other events have taken place since 1850. South Tropical Disturbances frequently start near to the preceding end of the Great Red Spot. Frequently a dark disturbance affecting the STB and the south edge of the SEB is seen and both belts may become dislocated into a South Tropical streak. Typically, the disturbance may last for a year or more.

The NEB, like the SEB, has sometimes experienced a fading, followed by a revival three years later. However, NEB fadings are rather less dramatic than NTB fadings, one of which is in progress as this chapter is being typed. The NTB, a much thinner belt than the SEB or NEB, can disappear entirely, and often fades on a ten-year basis. Occasionally, stretches of the NNTB have also faded at the same time.

There is always something for the Jupiter observer to monitor.

Once a new Jupiter imager has some experience, he or she may want to contribute some observations to a national or international body of observers, where their images can be measured. The Appendix contains a list of websites where such people can be contacted. Undoubtedly, the BAA's John Rogers is the expert on Jovian weather systems and we are lucky to have him as the Jupiter Section Director, but other organizations, such as the Association of Lunar and Planetary Observers (ALPO) have experts, too.

Without doubt, the most dramatic event to have occurred to Jupiter was the impact of Comet Shoemaker-Levy 9 with the planet in July 1994. This was a few years before webcams appeared on the astronomy scene, but not before a few amateurs like Donald Parker of Florida had equipped themselves with fast downloading CCD cameras and were taking high-resolution images. Many amateurs and professionals (including myself) thought that the impact of a string of kilometer-sized comet nuclei hitting the giant, 143,000-kilometer-diameter planet would have no detectable effect. How wrong we were! Huge black marks, bigger than the Earth, appeared on Jupiter and it looked like a planet sporting black eyes! Dropping a single droplet of ink into a bath of water produces a similar effect: the difference in volume may be huge, but the ink makes an obvious mark.

Webcam Imaging the Giant Planet

Compared to Venus and Mars, Jupiter is not a planet with a high surface brightness. Yes, it is bright in the evening sky, but that is mainly by virtue of its physical size. However, it is much brighter than Saturn and, crucially, more than bright enough for a webcam to easily record, even with brief 1/10th second exposures that freeze the seeing. The biggest problem with Jupiter is the speed with which it rotates. As we saw earlier, the formula to calculate the 0.5 arc-second drift time window of the center of the disc when the planet spans 45 arc-seconds, is: $0.5/((3.14 \times 45)/590) =$ a mere 2.1 minutes. Unlike Saturn, whose globe is less than half the angular size and whose features are very subtle, we really do have a serious problem here in keeping to the time margin if we are imaging with filters. With a color webcam, changing filters is not necessary, but when aiming for the highest quality results, amateurs often use red, green, and blue filters. Splitting the 2.1 minutes into three gives just over 40 seconds for each run. However, I cannot stress just how tight things become in the cold, damp, and dark in such a frantic scenario. Typically, you may be observing through gaps in cloud. Your Schmidt-Cassegrain may be dewing up and a hairdryer is needed to de-dew it. Then you have to wait for the corrector plate (or Newtonian secondary mirror) to cool down again. As soon as things are at ambient they start dewing again. You are then expected to change filters every 40 seconds and may have to refocus and re-center the planet every time, as even knocking the tube slightly when you change the filters, will cause an alignment problem; remember, the field of view may only be an arc-minute wide. Unless you have a very slick motorized filter wheel and a rigid telescope mounting, changing filters three times within a 2-minute window can be very fraught indeed. Jupiter shows considerable detail in an infrared filter (700–1000 nanometers) and one option with this planet, especially when low down, is to just take monochrome IR images, as shown in Figure 13.10. Specialist imagers like Antonio Cidadao often concentrate on narrow-band IR Jupiter imaging at the methane band wavelengths of 619, 727, or 890 nanometers; the 890 nanometer band is where methane gas absorbs light the most and is the most important region to study. With photographic film this spectral region was invisible to the amateur.

Another ingenious option, as frequently practiced by Damian Peach, and also by this author, is to just image Jupiter through infrared/red and blue filters and synthesize green by adding the red and blue channels together (see Figure 13.11).

Figure 13.10. Images taken in the infra-red (700–1000 nm) by Antonio Cidadao with a 35-cm SCT from Portugal. Opposite sides of the planet are shown with the Great Red Spot on alternate limbs. Note the fine details visible at this wavelength. Image: A. Cidadao.

Figure 13.11. When Jupiter is low down and blurred, good images can still be obtained with just two filters: an IR filter and a blue filter. The diagram shows infra-red (700–900-nm) and blue filtered images of Jupiter taken by the author on March 28, 2005. A green image synthesized by adding infrared and blue data is shown in the top right. The lower image is an LRGB image where 50% of the luminance has been derived from the IR image (and the other 50% from R,G,B.) The RGB color info is from the IR, synthesized green, and blue images. Images stacked and sharpened in Registax and LRGB combined in Maxim DL. Image: M. Mobberley.

Remarkably, this does work well, and the red-filtered image can produce some very fine detail, too. However, correcting the color balance to make the planet look natural can be a major challenge. In the infrared, Jupiter's moons appear very bright and so much redder than normal in the color image. Unfortunately, although seeing and contrast are improved substantially in I band there are a couple of instances where I band images lose out. The brown NEB barges do not show up well at infrared wavelengths and neither do some of the orange features like the NNTB, because these features are not as dark at IR wavelengths as they are in the red and green. The synthesized green technique is a very useful one, especially for a low-altitude Jupiter where seeing and contrast are normally poor. However, a true green filter really brings out the best contrast in the NEB and gives it the deep brick red color that simulated green fails to capture.

Ironically, when the planet is far away from opposition, or conditions are poor, you have a larger time window. If Jupiter has just emerged from solar conjunction it may only be 30 arc-seconds across, in which case, with the same 0.5 arc-second drift limit, you have three minutes to collect the images. If Jupiter is low down and blurred and there is no hope of getting sharp images, even a four-minute imaging run can produce a very pleasing image, capturing all the detail conditions allow. If your color image is the result of a red/infrared luminance and a blue image, with green synthesized, you only really need to worry about the red AVI being two minutes long. The human eye only notices luminance sharpness anyway and when aligning the blue/synthesized green with the red detail on the planet (i.e., aligning the detail, not the limb), the alignment error will only affect the planetary limb. An alternative solution to swapping filters is to swap webcams; one can be used for luminance (B&W) and one for color (a standard commercial webcam). If you have a Barlow or a Powermate on each, fine . . . just remove one and slide the other in. But there are problems even with this approach. The B&W and color webcams will invariably have slightly different focus positions. This, in turn, can mean a slightly different focal length and a slightly different image size for the B&W and color information. Another problem with this approach is that PC's and software can lock up when switching from one USB webcam to another. The last thing you want to have to do in a two- or three-minute imaging window is to reboot the PC! If you have a large-aperture telescope and a site with frequently good seeing, a standard color webcam can produce good, strong, high signal-to-noise Jupiter images at exposures less than the usual 1/10th of a second. Remember, it is a good plan to keep to 10 frames per second to avoid compressing and corrupting the image with a USB 1.1 webcam, but reducing the exposure time will still give you a shorter exposure (even if the actual exposure time indicated is not reliable). In Florida, Don Parker has taken spectacular Jupiter images with a ToUcam Pro webcam at a relatively bright (f/14) on his 0.4 meter f/16 Newtonian. Florida is a site renowned for its good seeing. Reducing the focal length may lose you some resolution on the best nights, but on mediocre nights it will dramatically improve the finding and ease-of-centering of the planet.

Being a gaseous world, Jupiter will never look good on the raw webcam output unless you are imaging in the infrared. If you are using an IR filter, the IR image should be focused first as it will be the sharpest and you can usually ignore refocusing when the blue (and optional green, if green is not being synthesized) filters are used. Two other advantages of not using a green filter and, instead, synthesizing green as an average of "IR + blue" are as follows. Firstly, Jupiter is usually a similar brightness in the near infrared to the blue, so you do not waste time turning the webcam gain down when the bright green image comes up. Secondly, certain features, especially the moons, are very bright in the IR image and very dim in the blue. If green is synthesized from bright Moon red and dark Moon blue, the resulting satellite images look orangey and not too bright. If a real green is used in the mix, the moons usually end up looking bright red. Remember, when using the LRGB method, the luminance image does not need to be the 100% luminance contributor. An LRGB image composed from 100% IR luminance and with its red color derived from the IR data, too, may look very artificial. In packages like Maxim DL, the luminance contribution can be altered in the LRGB mix. The most natural ratio seems to be one where the luminance weighting is 50%, i.e., 50% from the IR and 50% from the normal RGB values, even when the R is really IR and the green is synthesized. So, in a package like Maxim DL you would simply set red as IR, blue as blue, green as the synthesized IR-blue average image, and luminance as IR with a 50% weighting. Basically, you need to do plenty of experiments to see what looks the most natural.

Only the moons and their shadows will ever look sharp in a commercial color webcam; the rest of the planet will look drab and washed out as you image it. It is a good idea to reduce the gamma value in the webcam properties window when imaging Jupiter. A high gamma may make Jupiter's zones too bright and they will wash out. We want mid-range brightness values to be slightly dimmer than normal with Jupiter so that the zones will not swamp the fine detail, such as bluish projections and festoons entering the bright equatorial zone. This swamping of detail can partially be compensated for later in other packages, too, if the original webcam gamma is untouched. In this case, a gamma adjustment from 1.0 to around 0.6 seems to work well. With a basic color webcam like the ToUcam Pro, Jupiter seems to come out with rather a strange greenish tint. Tweaking the color to increase the red and decrease the green seems to work well with the giant planet. Alternatively, use the auto white-balance tool in Photoshop as a good starting point. A final tweak worth noting with any planet, especially Jupiter or Saturn, is, as a last stage, to adjust the brightness such that the very brightest part of the planet's globe (i.e., the equator/meridian central region) is almost whiting out. This may seem counterintuitive when considering what I have just said about gamma, but it will ensure that your final image has a good dynamic range.

With Jupiter, the problem for all photographers used to be the planet's limb-darkening. It is not really a problem with the powerful image-processing tools we have today, but making the brightest parts of the image reach 100% will ensure that Jupiter's limb does not make the planet look smaller than it is. This is essential if the images are going to be measured by students of the planet. If you are submitting your images for analysis, make sure you know exactly what time the mid-point of your exposure was. This can easily be determined using Windows Explorer and right clicking on the AVI file name and then selecting properties. The "created" and "modified" dates will tell you the start and end times of your webcam run. Bear in mind that it is worth checking that your PC clock is accurate when you do this. If a PC has locked up, and needed re-booting in recent weeks, the clock can be very inaccurate.

Animated GIFs

Images of the giant planet can be particularly spectacular if they are made into an animation. Because Jupiter rotates in under 10 hours and because it is big and detailed, a set of images taken half an hour apart over a period of several hours can make quite an impressive "jerky movie" of the planet rotating. Over a few hours, foreshortened features, initially just detectable coming into view on the morning terminator will be disappearing from view on the evening terminator. If a Jovian satellite (and/or its shadow) can be included in the animation, so much the better. The GIF image format lends itself very well to this type of work. GIF files are lossless (i.e., the image preserves all of the information even though it is compressed) and when originally designed by Compuserve, well before the modern era, the design thoughtfully included the facility for animations, i.e., numerous images within one file. Whereas jpeg compression works using a mathematical transform that makes the image more blocky and corrupted as the compression is increased (compression should be set to zero for planetary images), the GIF format simply encodes the data without corruption, but using an economic digital "shorthand." If a row of pixels in a GIF image is all the same brightness and color, the file size is dramatically reduced, hence the preference for GIF in block diagrams of a few specific colors. GIF images are at best 8 bit though, so jpegs converted to GIFs can look artificially contoured at times. Nevertheless, the animated GIF format is a superb way of producing jerky movies which can be e-mailed without clogging the recipient's mailbox. Virtually all modern image-processing software includes an animation package. Animated jpegs can also be produced using web page design techniques, but producing animated GIFs is far simpler at the time of writing. Individual images from a GIF animation by Jamie Cooper are shown in Figure 13.12. Jasc's Paint Shop Pro has an animator (Anim.exe) and Photoshop can produce animations by making each image a different "layer."

Figure 13.12. Even when Jupiter is at a low altitude, impressive animations can be compiled from images taken through infra-red filters in the 700–1000 nm region. These eight images were taken in one night in April 2005 and compiled into an animated Gif. Image: Jamie Cooper.

Strip Maps

Another method for visualising the appearance of a planet as more of it rotates into view is the strip map technique, i.e., a long image strip with a vertical height between, say, 60 degrees south and 60 degrees north and a horizontal width of up to 360 degrees. An example, by Damian Peach, is shown in Figure 13.13. Obviously, it is impractical to include the polar regions in such a strip as the higher latitudes get increasingly distorted if they are represented as a strip map. Even at 60 degrees latitude the circumference of a globe is only half that at the equator. Astronomical software packages such as Christian Buil's IRIS can be used to convert images of planetary discs into strip maps.

Finally, I would like to end this chapter with one of the most stunning images of Jupiter I have ever seen. Figure 13.14 was taken by Damian Peach using a 235-mm aperture SCT. At first glance it looks like a Hubble space telescope image. Amazing!

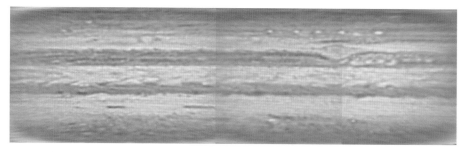

Figure 13.13. A stripmap of Jupiter covering 360 degrees of longitude, compiled from images taken over two five-hour periods on the nights of January 28 and 29, 2003. Celestron 11 image with a ToUcam Pro webcam. Image: Damian Peach.

Figure 13.14. This absolutely stunning image of Jupiter was secured by Damian Peach, from Barbados, on April 23, 2005. A Celestron C9.25 and Lumenera video camera were used for this composite of 3 × 1,000 frames at 17 frames per second.

CHAPTER FOURTEEN

Imaging Saturn

Without a doubt, Saturn is the *must see* planet that every one should view, through a telescope, in their lifetime. Most people, at some time, do actually look at the Moon, either with binoculars or a small telescope. However, to the naked eye, Saturn just appears as a bright star and is not obviously a planet. So unless you know where it is, you are stuffed! However, once you find it, the sight is mesmerizing. On first glance it seems just so incredible that a planet can have rings actually going round it: like something out of a science fiction film!

Saturn orbits the Sun at an average distance of 1427 million kilometers or roughly 9.4 times the distance of the Earth from the Sun. A photon of light takes 8.3 minutes to travel from the Sun to the Earth, but 80 minutes to get to distant Saturn. Even when Saturn is at its closest to us its light has taken over an hour to reach us.

Although the globe of Jupiter (equatorial diameter 142,880 kilometers) is almost 20% larger than the globe of Saturn (equatorial diameter 120,536 kilometers), Saturn's visible ring system spans an incredible 274,000 kilometers, that is 70% of the distance between the Earth and our own Moon. Put another way, it would take just under a second for light to travel across the ring system. Saturn's rings are the largest apparently single structure orbiting the Sun, although, in reality they are not a solid structure at all.

Saturn takes 29.4 years to orbit the Sun and during that time we view the northern hemisphere and the northern face of the rings and then the southern hemisphere and the southern face of the rings. This, of course, is because Saturn's rotation axis is tilted with respect to the plane of the solar system (called the ecliptic). The tilt is 26° 44′ or nearly 27° and thanks to this we are able to have a glorious view of the ring system. If Saturn had the relatively low axial tilt of Jupiter (a mere 3°) we would be denied the most spectacular sight through a telescope: Saturn with its rings wide open. Needless to say, as Saturn moves around in its orbit and the rings appear to close up, there comes a point where the rings are seen edge-on. In fact, a sequence of events takes place, because, while the Sun passes through the ring-plane in one go, illuminating first one side of the rings and then the other, the Earth, moving around in its orbit can peep just above and just below the ring plane several times before the rings really start to open up again. The sequence of events witnessed can be fascinating, although, since spacecraft visited Saturn, there is (probably) little scientific value to be derived. The tilt of Saturn's rings and the parallax created as Earth orbits the Sun is also the reason why Saturn's rings do not quite open and close smoothly; they can pause for a few months in the opening/closing process depending on where the Earth is relative to Saturn.

Because the orbit of Saturn is not a perfect circle, the time interval between successive edge-on events is not equal. We see the southern face of the rings for 13 years, 9 months and the northern face for 15 years, 9 months. While the southern ring face is fully exposed, Saturn reaches perihelion in its orbit and comes to opposition at a high northerly declination, as occurred on December 31, 2003. That is good news for high northern latitude observers like me. The rings will next be edge-on (with respect to the Sun) on August 10, 2009, May 5, 2025, and January 22, 2039. Saturn comes to opposition two weeks later every year.

The Saturnian Moons

Unlike Jupiter, Saturn only has one really large satellite. That Moon is, of course, Titan. Titan is the only planetary satellite to have an atmosphere and it is the second largest moon in the solar system: only Jupiter's moon Ganymede is larger. Titan has a diameter of 5150 kilometers, whereas Ganymede is a whopping 5,268 kilometers across. Surprisingly, Mercury, a planet in its own right is only 4,878 kilometers across, smaller than either, and Pluto is tiny, at 2,324 kilometers. Titan orbits Saturn every 16 days (a very similar revolution period to Callisto around Jupiter) but because the axial tilt of Saturn is almost 27 degrees, seeing Titan actually cross the disc of Saturn and seeing its shadow cross Saturn is an almost once-in-a-lifetime event. An image of such a rare event is something to be highly prized, so get ready for 2009! During the 1995/1996 edgewise presentation I identified a scant five nights when Titan, its shadow, or both would cross the disc of the planet, with the planet at a decent altitude in the U.K. night sky. Needless to say, all five nights were cloudy!

From Earth, Titan's disc, large though the satellite is, only spans a tiny 0.8 arc-seconds. Although amateur astronomers using webcams can resolve some surface markings on the large Galilean satellites of Jupiter, I have yet to be convinced that any real features can be resolved on Titan. Being a low-contrast, gaseous disc makes any claims to have resolved features impossible to prove. Nevertheless, the Spanish observer José Comas Sola was the first to realize that Titan had an atmosphere, when he suspected the moon exhibited limb darkening, in 1908.

But where Saturn has only one major 8th magnitude satellite, compared to Jupiter's four 5th magnitude moons, Saturn can boast many more fainter moons within visual and webcam range. Table 14.1 lists all the Saturnian moons within the visual/webcam range of an amateur telescope of 25–35-cm aperture. I have

Table 14.1 The Saturnian moons

Moon	Orbit Period	Distance	Diameter	Magnitude
Titan	15.95 days	1.22×10^6 Km	5,150 km.	8.3
Rhea	4.52 days	0.53	1,528	9.7
Tethys	1.89 days	0.29	1,058	10.2
Dione	2.74 days	0.38	1,120	10.4
Iapetus	79.3 days	3.56	1,460	10.2–11.9
Enceladus	1.37 days	0.28	~500	11.7
Mimas	12.9 days	0.19	~400	12.9
Hyperion	21.3 days	1.48	~300	14.2

listed them in descending brightness with the opposition magnitude being stated. Distance equals the distance from Saturn in millions of kilometers.

Iapetus has such a variable magnitude simply because one hemisphere is bright and the other is dark, thus, as its axial rotation is locked into the same period as its rotation it appears five times dimmer when east of Saturn.

I would like to digress slightly at this point into the subject of webcam sensitivity and limiting magnitude, as the question often arises when imaging Saturn and its moons. When imaging planets at long focal lengths, with a webcam, it might be imagined that the exposures (typically 1/10th second) were far too short to reveal faint satellites. It therefore comes as a bit of a shock, when the image brightness and contrast is adjusted, to see small dots emerging. However, this should not come as too much of a surprise, because just think of what the human eye can detect. In fact, the eye versus the CCD makes an interesting discussion point on its own. The retina's averted vision detectors, known as the rods, are extremely sensitive to light, although their resolution is quite poor. In comparison, the webcam pixels have fine resolution, but suffer from readout noise at the end of every 1/10th second exposure. The webcam does have two big advantages though: it is not "dazzled" by the presence of a nearby planet, plus, thousands of frames can be stacked to reduce noise. In fact, when a webcam stack of thousands of frames is pitted against the human eye, the outcome is surprisingly similar at long focal lengths. At f/30 or so, even a commercial color webcam on a 250-mm aperture will record Rhea, Tethys, Dione, and Enceladus on a contrast stretched image when atmospheric transparency is good. In other words, you can record stars and moons down to 11th magnitude without too much trouble. Of course, Tethys and Dione are the most frequently recorded moons on CCD frames of Saturn as not only are they relatively bright, they are close to Saturn, too. Titan, despite being the brightest Moon, is too far out to creep into planetary frames (except when the rings are edge on) although bright Rhea is often captured. Enceladus is right on the magnitude limit and Iapetus is far too distant from Saturn too usually enter the field. Lowering the f-ratio to 10 and using a B&W webcam can dramatically increase the number of satellites captured. At f/10 the wider, faster field captures Mimas as well as field stars down to 13th magnitude. The ability to record all these moons should not be surprising, as Saturn itself has a very low surface brightness. At 9.4 times the distance of the Earth from the Sun, light levels at Saturn are some 90 times dimmer than we enjoy and the brightness of the globe is only roughly magnitude 7 per square arc-second. From time to time, reasonably bright stars (i.e., 10th magnitude and brighter) are occulted by Saturn and its ring system. Amateur astronomers have had some success in imaging these events in recent years and the subsequent animations produced are fascinating to watch, as the stars are dimmed by the rings, disappear, and then wink back on in the Cassini division! Unfortunately, cloud, once again, can wreck all the plans to secure images of such time-critical events. If such a star appears very faint, dropping back to f/10, or increasing the webcam exposure to 1/5th second will ensure the star is recorded with a better signal-to-noise ratio.

Saturn's Weather and the Great White Spots

To the beginner, Saturn's globe appears fairly featureless when viewed through the eyepiece. Although Saturn does have equatorial, tropical, and temperate belts and zones (Figure 14.1), there is nothing on the disc that stands comparison with the Jovian Great Red Spot or even the giant planet's North and South Equatorial Belts. Undoubtedly, the huge amount of internal heat produced by Jupiter, combined with its closer proximity to the Sun, is responsible. Heat is the driving force behind weather systems and the ammonia crystals in Saturn's atmosphere form at higher levels, tending to give the planet its relatively featureless appearance. Features more than 3,000 kilometers in diameter are very rare in Saturn's atmosphere. However, an interesting phenomenon is that when the rings are near their maximum tilt with respect to the Sun, either the northern or southern hemispheres lie in the shadow of the rings. This cools the shadowed hemisphere and, when it reemerges into sunlight, the colder hemisphere often has a bluish tint, similar to that of colder Uranus and Neptune.

Despite the initially bland appearance of the planet, the experienced visual observer will easily spot some of the more distinctive features revealed in the best webcam images. Saturn's equatorial zone, like that of Jupiter, has a bright, creamy

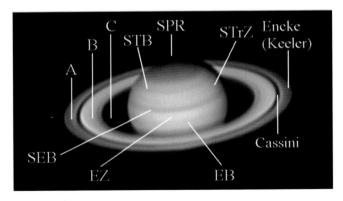

Figure 14.1. Saturn's rings, divisions, and belt nomenclature. Webcam image and diagram: M. Mobberley.

appearance compared to areas at higher latitudes. Darker, brownish, SEB and NEB belts, either side of the Equatorial zone are visible at higher latitudes, but these in no way compare to the Jovian SEB and NEB. However, with webcams and modern image processing techniques, a wealth of subtle detail can be revealed on the best nights by the best observers. The polar regions are especially interesting when viewed by the Hubble space telescope (Figure 14.2) or the world's most advanced amateur astronomers. At the precise position of the poles, within the south or north polar regions (SPR and NPR) a tiny dark circle can be resolved, like a bulls-eye on a dartboard. This is usually surrounded by concentric dark bands of various subtle colorations. In the last few years the whole polar region of the southern hemisphere has sometimes taken on a greenish hue, but, closer to the pole, dark red/brown, dark blue, and even dark yellow collars have been seen. The images of Damian Peach have revealed these colors in webcam sequences taken from 2002 to 2005 and consultation with professional astronomers using Hubble have confirmed that these colors are real and do change on a short timescale. As a rough generalization, in 2002/2003 the main polar cap collar had a distinctly greenish hue, but in 2004/2005 it was mainly an orange/red color. However, in 1999 this SPC collar was a really deep red color, which faded out, and became pale blue in 2000. Prior to the webcam revolution only Hubble regularly revealed these features.

According to professional astronomers, such color changes in Saturn's atmosphere are most probably due to small changes in the size of the aerosols at upper levels (sometimes in the stratospheric haze at pressures around 10 mbar, sometimes in the upper level of the tropospheric haze at the 70 mbar level). A size change from .5 to 1.5 microns at some latitudes can reproduce the color variability. In some cases it is also possible to have changes in composition (in particular in the stratosphere, due to photochemistry or polar particle bombardment), that change the refractive index of the particles.

Despite all this talk of subtle Saturnian features, now and again Saturn produces large storms in the equatorial zone that become very obvious even in small telescopes. These "Great White Spots" can be dramatic, brightening the entire Equatorial Zone all of the way around the planet. Major white spots have been observed in 1876, 1903, 1933, 1960, and 1990, i.e., at intervals very close to Saturn's orbital period around the Sun, of 29.4 years. On this basis the next major white

Figure 14.2. Saturn imaged by the Hubble space telescope on March 22, 2004. An image processed to supposedly show the natural colors of the planet. Image: NASA STScI/ESA.

spot might be expected in 2020. Without a doubt the most highly publicized Great White Spot was that of 1933. Not only was it spectacular, its discovery was made by a famous British stage and screen comedian, Will Hay, on August 3rd of that year! The spot, shown in Figure 14.3, remained visible for six weeks.

In recent years, from observations by Hubble and by leading amateur astronomers, it has become apparent that relatively small white spots, typically 2,000 to 3,000 kilometers in diameter can be detected by amateur telescopes of only 25 cm in aperture. Darker spots have also been imaged at higher latitudes, in the polar regions. Such spots are very low contrast features though and only the best amateur images taken under near-perfect seeing conditions can reveal them.

Saturn through the Webcam

Saturn is the most challenging and, arguably, the most rewarding target for the webcam imager. The biggest problem posed by the ringed planet is simply its low surface brightness. Saturn's globe is 16 times fainter than the globe of Mars and a third as bright as Jupiter. With the maximum available AVI video exposure being 1/5th of a second (5 frames per second at, confusingly, the 1/25th setting on the Philips driver settings) and 1/10th being more desirable (to freeze the seeing) the raw Saturn image is always going to be a faint and noisy affair at focal ratios of f/25, f/30, and higher. Of course, a larger aperture, permitting an f-ratio of 20 or smaller will improve matters, but larger is not always better. As we learned at the start of this book, telescopes over 25 cm in size have considerable thermal mass and some take hours to cool to the night air; especially those with mirrors more than about 45-mm thick. Large optics in amateur hands rarely have the accuracy of smaller quality optics and they are also harder to operate and set up, except when installed in a permanent observatory. In the latter case, massive observatory domes cause further thermal problems. In recent years, amateurs using Schmidt-Cassegrains, Newtonians, and Maksutovs in the 23- to 30-cm range have taken Saturn images that compare with any taken with much larger ground-based instruments. Telescopes above 35 cm aperture often seem to have insurmountable thermal problems unless they are equipped with fan cooling or are used at dawn,

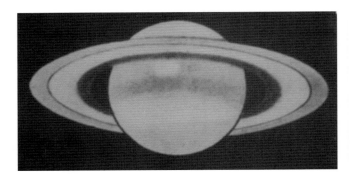

Figure 14.3. The Great White Spot on Saturn discovered by stage and screen comedian Will Hay on August 3, 1933. This drawing by Hay is from Volume 44 of the *BAA Journal*.

when everything is in thermal equilibrium. While excellent deep red lunar images have been taken with amateur telescopes of 45-, 50-, or 60-cm aperture, the same cannot be said for color planetary images where the best 25 cm ones are rarely beaten. Another factor here of course is the maximum aperture that the atmosphere will allow. As an observer of over 30 years experience, using telescopes from 22 to 49 cm, I would say that, on 95% of clear nights, a 25-cm aperture will give you as good an image as you can ever expect from the U.K. and a larger aperture will frequently give you a myriad of overlapping images, in other words, a larger aperture can be worse! Of course, there are exceptions to every rule though, and the results of Florida's Don Parker and Japan's Isao Miyazaki, with 0.4-m reflectors at superb locations, prove that high-quality instruments, in the right hands, can deliver the goods, even if they do not outperform 25-cm apertures. As I mentioned much earlier, Florida's Maurizio Di Sciullo is, at the time of writing, constructing a 36-cm planetary Newtonian with a water-cooled primary to try to overcome the inherent cooling problems of such a large glass blank.

So, assuming we are using an aperture of around 25 cm, how can we salvage a result from faint Saturn? The standard color webcams like the ToUcam Pro can still produce spectacular Saturn images in the hands of an expert like Damian Peach, especially when the planet is at high altitude and color dispersion is minimized. One of Damian's ToUcam images is shown in Figure 14.4. One option is to use a sensitive B&W monochrome webcam such as the ATiK 1HS to acquire an unfiltered luminance image and then to used red, green, and blue filters to acquire the color. Or, the red and green images can be added to create a very good luminance image. Another option is to increase the webcam frame rate to its slowest rate of 5 frames per second.

Figure 14.4. Saturn, imaged on October 28, 2003, by Damian Peach, using a 280-mm aperture Celestron 11 SCT at f/30 and a ToUcam Pro webcam. 3000 thousand frames taken at 1/5th-second exposure (5 frames per second on the misleading 1/25th setting) were used in the final image stack. The raw stacked image is at the top and the final processed image below. Image: D. Peach.

As we have already seen, at 5 frames per second, most webcams will be exposing for 1/5th of a second in manual mode, regardless of the actual exposure time registered. Of course, in poor seeing, 1/5th of a second will not freeze the view. It is easy to get carried away when an image on the PC screen is faint and increase the webcam gain to 100%. Unfortunately, this will result in extremely noisy frames. Usually, a dim (but clearly visible) image will have a much better signal-to-noise. So try to keep the webcam gain to about 85% maximum. A related issue in this context is that very low light levels can confuse the auto color-balance settings in a commercial color webcam. Yet another possible factor relating to noise is, of course, the ambient temperature. There is not much that can be done about this, short of fan cooling or Peltier cooling the webcam chip. Fan cooling the webcam electronics may reduce the chip temperature by 6 or 7°C, thus halving electronic noise, but serious Peltier cooling introduces potential condensation problems, which, in turn, can lead to artifacts when processing the image. Also, it is readout noise, not thermal noise that dominates in webcam frames. Peltier-cooled webcams are only really necessary in long exposures or in hot tropical climates. Another alternative approach for Saturn is not to use a webcam at all. Modern cooled CCD cameras have fast USB download times and are extremely quantum efficient. They can also take exposures of much longer than 1/5th of a second but, again, this reduces the chance of being able to freeze the seeing. Of course, a webcam can cost you well under $100, whereas dedicated CCD cameras cost well over $1,000. Software such as SBIG's planet master can be used to select the best frames and stack them. Using a cooled CCD camera on Saturn you will end up stacking 100 or 200 frames, rather than thousands of noisier webcam frames. However, let us not get too pessimistic here. Truly excellent webcam pictures of Saturn have been secured with f-ratios as high as 40, as long as there are thousands of frames to stack. One such image, with a 235-mm aperture, is shown in Figure 14.5.

Figure 14.5. Amazingly, this superb image, taken on December 11, 2004, was made with a relatively modest aperture Celestron 9.25 (235-mm aperture) SCT working at f/40. Images taken through red, green, and blue filters with an ATiK 1HS webcam were used for this remarkable composite of several thousand frames. Image: Damian Peach.

Saturn's Time Window

Fortunately, Saturn does have one big advantage compared to Jupiter. The visible globe, at 20 arc-seconds in diameter, is less than half that of Jupiter at opposition. Thus, the planet takes longer to rotate before features exceed our 0.5 arc-second equatorial drift limit. The rotation period, at 10 hours, 14 minutes, is slightly longer too. Another factor here is that the Saturnian globe contains no high-contrast features. It will not really matter if features do drift a bit. Also, most Saturnian features are at relatively high latitudes, too; a spot at 60 degrees north or south will travel at half the speed across the planet's meridian as a spot at the equator. Subtle color changes take place in the polar regions every year and can be recorded by amateurs, as shown in Figure 14.6. With regard to image smearing with the planet's rotation, the simple fact is that, unless thousands of webcam frames are stacked together, the subtle, low-contrast Saturnian spots will never emerge from the noise anyway, at f/30, unless only red images are used, where the CCD is more sensitive and seeing generally better. To record Saturn's subtle features it may well be necessary to drift beyond the self-imposed limit of 0.5 arc-seconds. To record the subtle color changes in Saturn's polar regions it is perfectly OK to image for 15 minutes or so. If the equator drifts by over 1 arc-second, but there are no features in it, who cares!

Applying our well-used formula, and inserting 20 arc-seconds for the opposition diameter of the planet and 614 minutes for the rotation period gives us:

$$0.5/((3.14 \times 20)/614) = 4.9 \text{ minutes}$$

In practice, increasing this value to around 6 minutes will not produce any noticeable smearing on high-latitude spots. A 6-minute time window is obviously a far more manageable time span for changing filters than the 2-minute time span we suffer with Jupiter. At 10 frames per second, up to 3,600 frames can be exposed and even if almost half of these are rejected for poor quality, an image consisting of 2,000 frames will have quite a smooth appearance. As I have said, when there are no spots on the disc, or when seeing is not excellent, there is no reason why a 10- or 15-minute run should not be stacked. Such a composite will produce a very smooth result, giving a high-quality, if largely featureless, image. However, for images subjected to scientific analysis the precise exposure details should always

Figure 14.6. Subtle changes in the color of Saturn's south polar regions over three observing seasons can be recorded by amateur observers. Image: Damian Peach.

be recorded and supplied to the analysing body. Although Saturn appears sharpest in red and green frames when using filtered imaging, the blue frames are the highest contrast as regards revealing the darker belts. Reducing the blue content of the image will increase the sharpness but also decrease the contrast of the globe, which will tend to become featureless except at the poles.

Saturn's Rings

As well as Saturn's rings and the Cassini division providing an excellent sharp edge on which to focus, they also act as a great seeing indicator and can be used to check the color balance of an image. Let us look at the ring structure in more detail.

The Pioneer 11, Voyager 1 and 2, and Cassini spacecraft all showed that the Saturnian ring system is highly complex. Fortunately, the view through the amateur's telescope is a lot simpler and basically boils down to a system of three rings, called, not surprisingly, A, B and C. Saturn's A ring is the outer one and has a grayish or grayish-blue cast. The A ring, as we saw earlier, has a diameter of 274,000 kilometers and a width of 14,600 kilometers. Near the outer edge of Ring A, some 90% of the way from inside to outside edge is the black gap called the Encke division. It has a width of only 325 kilometers and so subtends only 1/20th of an arcsecond as seen from the Earth! However, because it is such a dark feature, it can be detected, as a contrast drop in ring A, with instruments as small as 15 or 20-cm in aperture, despite being well below their stellar resolving power. The Encke division is, undoubtedly, the ultimate planetary resolution goal for any amateur imager and can only be seen or imaged when seeing conditions are close to perfection. However, under such situations it can even be detected, as the merest dark sliver, on single raw frames. Before the CCD/webcam imaging era, visual observers rarely, if ever, spotted the Encke division. Indeed, prior to Pioneer 11's flyby of the planet in September 1979 the exact position and even the existence/permanency of the feature was disputed, despite the fact that James Keeler had sketched the position accurately, using the 36-inch Lick Refractor, in January 1888. When Saturn's rings are wide open and conditions are close to perfection the Encke division can be imaged almost all around the rings with 25-cm apertures. However, in practice, even stunning amateur images rarely capture the Encke division far from the ring ansae (the east and west tips) and as the rings are now narrowing the Encke division will just become harder to resolve, even at those points. To my knowledge the first amateur image ever showing the Encke division was obtained in 1998 by the French imager Thierry Legault, using a 30-cm Meade LX200 and a Hi-Sis 22 CCD camera. Prior to that, the only images showing the feature clearly from Earth seem to be ones taken with the 1-meter Pic du Midi Cassegrain in the Pyrenees and 1.54-meter Catalina telescopes in Arizona. However, since the webcam era, the feature has been resolved annually in the very best amateur images. It is interesting to note that, quite often, the feature will only be recorded at one ansae and not both. This may well be because the feature is so thin that if the CCD chip is not perfectly flat in its base, one side may be in focus and the other side fractionally out-of-focus, rather than a physical difference in the Encke division width. Imaging the Moon with the same set-up can identify problems of this nature.

The Encke division, which is incredibly thin and dark, should not be confused with the Encke minimum, which is more of a perceived gradual contrast feature across the A ring thickness, making the outer half of the ring appear slightly darker than the inner and giving the illusion of a division in the A ring center.

It is important to note that spurious ring divisions can easily occur when stacking thousands of images of Saturn together. Think, for a moment about what is happening in less than perfect seeing. The atmosphere not only blurs the planet, it physically distorts it, too. The most popular stacking software, Registax, tries to cope with this by rejecting the most distorted images, but, the fact remains that the boundary between Saturn's A ring and the night sky is a high-contrast feature. If the position shifts too much, artificial ring divisions can appear.

Moving in from the gray outer ring we come to the much wider Cassini division. This chasm is 4,700 kilometers wide and so spans almost 0.8 arc-seconds at opposition. Even in poor seeing the Cassini division is usually easy to spot when the rings are wide open, but if seeing is appallingly bad, it too can disappear in small telescopes! These days, once thousands of webcam frames are stacked, on a reasonable night, the Cassini division can easily be traced all the way around the planet. However, by the time you purchase this book, the rings will be closing noticeably and, as 2009 approaches, even the Cassini division will become a challenge.

Ring B is Saturn's widest ring, at 25,500 kilometers from inner to outer edge. Unlike the A ring it appears to be almost colorless (a pure white or gray) and is, therefore, an excellent check on how good the color balance in the final image is. Ring B has nothing like the Encke division within it but, it does gradually darken as you move in toward Saturn. At the point where it almost disappears it merges into the illusive C or "Crepe" ring.

The Crepe ring is a feature that even today's CCD technology struggles to record. Not because it is narrow, it is not. The problem lies in the faint, ghostly nature of the C ring. Increasing the brightness and contrast of a webcam or CCD image will reveal the transparent Crepe ring through which the globe of Saturn is clearly visible. However, the dynamic range of even today's electronic detectors cannot quite match the human eye's ability to stare at the globe of Saturn and the Crepe ring and see them both together. Increasing the brightness and contrast of a webcam image to show the Crepe ring will saturate the planet's globe, but reducing the brightness and contrast to normal levels will sink the Crepe ring down into the blackness. In most amateur webcam images, the Crepe ring is only well seen where it crosses in front of the globe, and the planet can clearly be seen through it. On January 13, 2005, I was lucky enough to observe Saturn within a few minutes of opposition (i.e., Saturn opposite the Sun in the sky). The brightness of the rings was dramatic, as always happens that close to opposition. My colleagues Damian Peach and Dave Tyler were also imaging that night and we were all amazed at the increase in the ring brightness and the apparent dullness of the globe by comparison (see Figure 14.7). This is called the Seeliger effect and occurs partly because the individual ring particles are not casting shadows on top of each other. But, most noticeable of all was the obvious presence of the Crepe ring; normally ghostly, it was very obvious. Damian and Dave produced a stunning composite image from opposition night, consisting of 9,500 stacked frames.

Figure 14.7. Possibly the best amateur image of Saturn ever taken? Damian Peach and Dave Tyler, using Celestron 9.25 and Celestron 11 SCTs (respectively) and living only a mile apart, secured excellent images of the ringed planet on January 13, 2005, within a few minutes of precise opposition! Note the extreme brightness of the rings and the relatively dull globe at precise opposition. Both observers exposed thousands of red, green, and blue frames in 15-minute time windows and the resulting composite consists of 9,500 webcam frames! Image: D. Peach and D. Tyler.

Despite the faintness of Saturn's globe it can, remarkably, be imaged in morning and evening twilight when conditions are often very stable as the atmospheric cooling is at a minimum. The planet can be located, with the aid of a GO TO system, or setting circles, shortly after sunset and imaging can take place with the Sun less than 10 degrees below the horizon. In such situations, in the evening, the red images are best acquired first and the blue last, as the twilight sky is mainly blue and will interfere with the blue frames if twilight is too bright. At dawn the blue images are best acquired first, and the red images last. Twilight images are often inevitable when the planet is far from opposition. On these occasions the shadow of the Saturnian globe is quite evident as it spreads over the rings behind the planet and off to one side. A twilight image almost three months after opposition is shown in Figure 14.8.

Finally, before we leave Saturn, let us remind ourselves once more, that at low f-ratios, the ringed planet's brighter moons can all be captured in a webcam frame, even with 0.1 second exposures. Figure 14.9, by Dave Tyler, is a clever composite of a low-gain imaging run to capture the planet and a high-gain imaging run to record the moons. Both AVIs were taken within a few minutes and then superimposed. As the rings close up in the next few years, the moons will tend to appear more frequently in our webcam images.

Figure 14.8. Saturn imaged nearly three months after opposition by Dave Tyler on April 2, 2005. The image was taken in a very calm seeing period in evening nautical twilight. Celestron 11 at f/40. ATiK 1HS webcam. Image: D. Tyler.

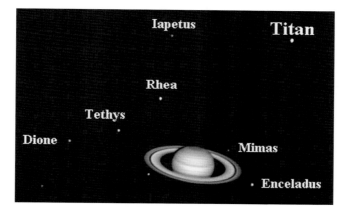

Figure 14.9. Saturn's moons can easily be captured with a webcam. In this f/10 composite by Dave Tyler (Celestron 11), seven of Saturn's moons and two field stars have been captured with individual exposures of only 0.1 seconds. The faintest Moon, Mimas, is only magnitude 12.9. Image: D. Tyler.

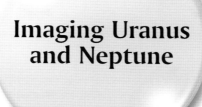

CHAPTER FIFTEEN

Imaging Uranus and Neptune

This chapter could not have been envisaged before the webcam era; it would have been considered ludicrous! Uranus and Neptune are in a different category to all the other planets I have mentioned (with the possible exception of Mercury) because they are just so tiny even through a quality amateur telescope. Not only that, but even compared to distant Saturn they are a *very* long way away from the Sun, and therefore very dim. The critical parameters are detailed in Table 15.1.

At a glance, we can see that even at an image scale of 0.2 arc-seconds per webcam pixel, Uranus currently spans less than 19 pixels and Neptune only spans 12! Surely, little can be done with these planets. Well, we must be realistic here. There is, indeed, little prospect of revealing much more than major, global weather upheavals, but who is to say that such events do not occur? The image by Christophe Pellier in Figure 15.1 hints that major atmospheric events would be detectable, even with a small aperture. Only one spacecraft has passed Uranus and Neptune at close range: Voyager 2 in January 1986 and August 1989, respectively. The Hubble Space Telescope has imaged both planets from time to time and, in recent years, the giant Keck telescope on Hawaii has obtained Hubble-quality images too, using advanced adaptive optics techniques.

As we have seen with Mars, a small planetary disc does have some "silver lining" advantages for the webcam imager. With the Earth's atmosphere imposing a resolution of 0.5 arc-seconds on the overwhelming majority of nights, if a planetary disc appears tiny, it can be allowed to rotate for tens of minutes before the rotation smear exceeds the atmospheric/telescopic resolution. This is a big advantage with planets as faint as Uranus and Neptune where a webcam will be working right on its limit. Uranus rotates in 17 hours, 14 minutes and Neptune rotates in 16 hours, 7 minutes. However, there is an interesting twist with Uranus. Neptune has an axial tilt of 28.8°, similar to Saturn, but Uranus spins virtually on its side! The axial tilt of Uranus is almost 98°, so, as this is more than a right angle, it is technically rotating

Table 15.1 Critical Parameters of Uranus and Neptune

	Mean Distance from Sun (millions km.)	Orbital Period (years)	Diameter (equatorial kms)	2006 Opposition Diameter (arc-seconds)
Uranus	2,870	84	51,118	3.7″
Neptune	4,500	165	50,538	2.4″

Figure 15.1. Uranus imaged with a modest 180-mm Newtonian and ATiK 1HS webcam on July 6, 2004. Image: Christophe Pellier.

backwards (like Venus, except much faster). So, at times, as Uranus moves around in its 84-year orbit, it will point its north pole towards us, then, 20 or so years later, its equator, then its south pole, then its equator again! Uranus' south pole was pointed at us in 1985. In 2030 we will see the north pole. So around the publication date of this book we are seeing both Uranus' hemispheres, much like for any other planet except, its axis will be horizontal as we look at it.

Applying our favorite formula for a maximum drift window to Uranus and Neptune gives us:

0.5″/((3.14 × 3.7″)/1034 minutes) = 44.5 minutes

and

0.5/((3.14 × 2.4″)/967 minutes) = 64.2 minutes

In other words, we have 45 and 62 minutes, respectively, in which to collect the frames before Uranus and Neptune's central features have drifted by half an arc-second. Imaging these planets will not be a rushed affair! Being realistic, Neptune is, perhaps, an object best left to the Hubble Space Telescope (see Figure 15.2). As for Pluto, well, that is without doubt, purely a Hubble target (Figure 15.3), even though it is possible to view it visually, as a dot, through a large amateur telescope.

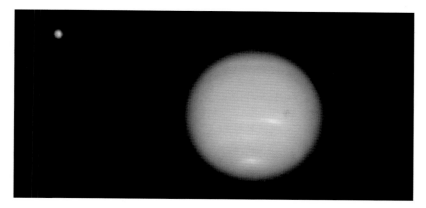

Figure 15.2. Neptune and it's large Moon Triton imaged by the Hubble space telescope. Image: NASA/STScl.

When taking images of objects that are as small as Uranus and Neptune, almost featureless, and so rarely imaged, it is impossible to verify the details that emerge after extreme image processing. The effects of atmospheric dispersion in Earth's atmosphere alone can cover most of the planet, making any features resolved somewhat speculative! To really unequivocally resolve details on Uranus and Neptune you have to image them with the Hubble Space Telescope or with ground-based telescopes with adaptive optics that can compensate for the effects of the Earth's atmosphere and are sited at the best locations on Earth.

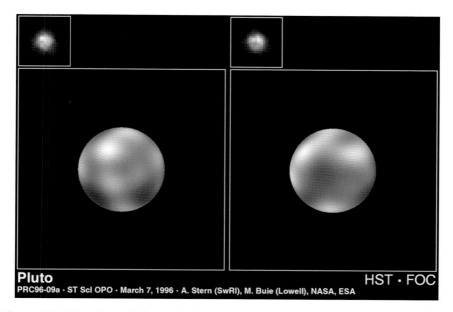

Pluto   HST · FOC
PRC96-09a · ST Scl OPO · March 7, 1996 · A. Stern (SwRI), M. Buie (Lowell), NASA, ESA

Figure 15.3. Two faces of Pluto and Charon imaged by the Hubble space telescope on March 7, 1996. Image: NASA/STScl.

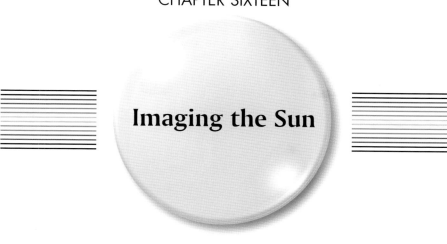

CHAPTER SIXTEEN

Imaging the Sun

When I started writing this book I had no plans to include a chapter on solar imaging. Why? Because the Sun is a dangerous object that should never, *ever* be observed either with the naked eye or through a telescope unless, in the latter case, the amateur astronomer is highly experienced and equipped with the right filters. Personally, although I would class myself as an experienced amateur and I do have the right filters, I still feel very nervous about observing the Sun, even with expensive equipment. Even experienced amateurs have made mistakes in the past and damaged their eyesight. The most common error is to imagine that a filter that dims the solar disk to a pleasing level is actually safe. DO NOT BE FOOLED! It is the infrared (heat) radiation that can destroy the retina, and that is invisible. However, despite my fears about solar observing I decided that, in the webcam era totally safe solar observing was possible: hence this chapter. The safest way to observe sunspots on the solar disc, without webcam equipment, is to project the image onto card, using a small refractor. A large cardboard shield can be used to prevent direct sunlight washing out the projected image. It is interesting to note that the vast majority of eyesight injuries, following events like solar eclipses, are due to people staring at the Sun through half closed eyes, in the misguided belief that this must be harmless. The lack of pain-generating nerves in the retina gives the illusion that you are just being dazzled and all is well . . . WRONG! **NEVER STARE AT THE SUN WITH THE NAKED EYE!**

Fortunately, the webcam, yet again, offers a superb way of imaging sunspots at high resolution but with absolutely no risk to the eyesight. In addition, a webcam is cheap, and even if it is damaged, it is not a big problem. Damian Peach, who is one of the world's leading planetary imagers, had an interesting experience when using a webcam to image the Sun. While packing the equipment away, he took the solar filter off first and then started to pack his PC away. When he next emerged from the house the webcam was totally ablaze! By removing the special solar filter the amount of heat and light falling on the webcam had increased by 100,000 times, setting fire to the webcam. However, as I pointed out to him, only the day before he had been imaging the Sun with a new $2,000 digital SLR. Maybe the webcam accident was a timely reminder of the Sun's power. Even when you are not observing visually accidents can happen. Telescope finders should be capped as well as the main instrument. There has been at least one documented case of a beard igniting after being set ablaze by light from the finder! Also, when you are packing equipment away and removing solar filters, point the instrument away from the Sun first. The heat entering even a small telescope is enough to melt a plastic secondary mirror holder or to make the glue on a Maksutov baffle tube go soggy and melt!

Large telescopes are not necessary for solar observing, and for two reasons. Firstly, the Sun is extremely bright, even when filtered, so light grasp is not an issue. Secondly, when the Sun is above the horizon, atmospheric seeing is at its worst, so a large aperture rarely gives any resolution advantage. A 10- or 12-cm refractor is the most that is needed and even an 8-cm instrument can get stunning results. The solar brightness leads to one big advantage, exposure times can be very brief to freeze the moments of good seeing. In addition, as there is no planetary rotation problem and Registax' alignment software will lock onto the shape of a sunspot, there is no time window restriction for imaging.

Solar Facts

The Sun is an absolutely massive body with an equatorial diameter of 1.392 million kilometers. Over 1.3 million Earths could fit inside the Sun and it would take a third of a million Earths to equal the Sun's weight! It is estimated that the Sun accounts for more than 99% of the mass of the solar system. The Sun rotates in 25.38 days with respect to the stars. As we orbit the Sun, we see it rotate in a slightly longer period, namely, 27.28 days. At our closest to the Sun (perihelion occurs around January 2), we are 147.1 million kilometers away. At our furthest (aphelion occurs around July 5) we are 152.1 million kilometers away. Put another way, we orbit the Sun at an average distance of 149.6 million kilometers ± 1.7%. Light takes 8.3 minutes to travel from the Sun to the Earth. The brilliant surface of the Sun is called the photosphere and the most notable features on this blinding yellow surface are the sunspots. Sunspots look dark because, at around 4000°C, they are cooler and dimmer than the 5000°C photosphere around them. However, if you could see them in isolation they would glow like an arc lamp. Sunspots are, primarily, the result of intense bipolar magnetic fields formed where concentrated fields emerge from the photosphere. If you place some iron filings on a sheet of thick card and move a powerful bar magnet under the card, the filings will move around in a very similar manner to high-resolution animations of sunspots evolving with time! Long-lived sunspot groups can rotate off the limb of the solar disc and emerge on the opposite limb, still in-tact, two weeks later. Indeed, the very largest spot groups survive for several solar rotations. The largest sunspot group recorded was that of April 1947, which covered an area of 18 billion square kilometers. In other words, it had an area almost 40 times larger than the surface area of the Earth! Such huge sunspot groups can span two or three arc-minutes in size, i.e., a larger angle than any planet spans as seen from Earth and equal to the angle subtended by the largest lunar craters. The Sun has an 11-year cycle of activity during which a lot of sunspot activity is observed at the start (solar maximum), followed by a virtually spot-free Sun (solar minimum) some 51/2 years later, and another maximum after 11 years. During total solar eclipses the very outer atmosphere of the Sun appears almost symmetrical at solar maximum, but highly east-west elongated at solar minimum. Huge solar prominences arching above the Sun can be seen at total eclipses or if you have expensive H-Alpha filters (read on). When a really major solar flare erupts on the Sun and charged particles head earthward in a CME (coronal mass ejection), the results can be awesome. If the mass hits the Earth's magnetic field (a rare event), a spectacular aurora can be seen even at temperate latitudes. The last solar maximum occurred in 2001/2002. The next should be in 2012/2013.

Solar Filters

The Sun is a unique target for the amateur astronomer. In every other field of astro-imaging more light is an advantage, but in solar imaging you just have to reduce the light and the heat coming through to your webcam. Even on the shortest webcam exposures the unfiltered Sun will not only saturate the image but it will definitely set fire to the webcam! A substantial proportion of the light and heat must be rejected by filtering. So which solar filters are the best to use? The first thing to say here is that choosing a solar filter for a telescope is a serious issue. I can remember a time when almost every cheap and nasty shopping mall refractor came supplied with a lethal solar filter. These horrors screwed into the eyepiece barrel, where all the heat concentrates, and they were liable to crack due to the heat falling on them. Thankfully, these nightmare filters appear to have become extinct and all modern solar filters are full aperture ones. By full aperture we mean that they fit at the very front of the telescope. With a refractor this means over the dew cap or the objective lens. Many telescope manufacturers supply full aperture solar filters that are specifically designed to tightly fit, without the risk of a gap, over their own instruments. The larger solar filter suppliers also sell filters designed to fit the most purchased telescopes, such as Meade SCT's refractors and ETX Maksutovs. A filter that fits tightly across the telescope aperture is essential. Off-axis solar filters are also available for large-aperture telescopes like Meade and Celestron's Schmidt-Cassegrains. With these filters most of the aperture is blocked completely, but a short, off-axis aperture (between the secondary mirror holder and the aperture rim) is filtered, thus turning a large Schmidt-Cassegrain into a smaller-aperture solar telescope.

The daytime sky, with the Sun heating the atmosphere, features the worst atmospheric turbulence that an amateur astronomer has to contend with: the very object being imaged is causing the turbulence that wrecks the view! However, the sheer brightness of the Sun means that ultrashort exposures can freeze the seeing effectively, and the two factors almost cancel out: almost, but not quite. Daytime resolution is still generally worse. The brighter the image, the shorter the exposure you can employ, and modern webcams allow exposures as short as 1/10,000th of a second. Choosing a solar filter that lets through more light than a safe visual filter can be a distinct advantage here, but such filters should never, *ever* be used visually. Solar filters are often given an ND rating. This stands for neutral density and is graded logarithmically. Thus, an ND 1 filter attenuates by a factor of 10; ND 2 = 100× attenuation; ND 3 = 1000× attenuation. The standard safe visual filters are ND5; in other words, only one part in 100,000 of the light gets through to the observer's eye, or, 99.999% of the light is blocked. The big six names in solar filter supplies are Baader, Thousand Oaks, Kendrick, Roger W. Tuthill, Orion Telescopes & Binoculars, and Celestron. Thousand Oaks and Kendrick will sell you ND4 filters, yielding a 10 times brighter image for photography/CCD/webcam use. Obviously these should never be used visually. In fact, Celestron and Kendrick use Baader's AstroSolar film for the basis of their full-aperture filters. Essentially, there are two types of safe solar filter in use by amateurs. The first type consists of an optically flat glass window. That type has a thin layer of aluminium deposited onto the glass (like the aluminium on your telescope mirror) but just thin enough to let 0.001% of the light through. The second type of filter is an ultra-thin aluminized mylar film. This is cheap, because it is far easier to make aluminized mylar than to mass produce optically flat glass discs and then aluminized them. It might be assumed that the glass filters are better, especially when you look at the rippling, crinkly, mylar filters. However, tests do not bear this out. The damage that will be done to the incoming light by an ultrathin sheet of mylar seems to be no worse than that inflicted by a much thicker, but flat, metal-on-glass filter, especially when compared to the damage wreaked by the Earth's atmosphere. From a safety viewpoint though, mylar film is easily creased and damaged and should always be carefully inspected before use. Of course, if you only ever observe the Sun with the webcam, safety may be less of a concern as a burning webcam can be replaced; a damaged retina cannot. With ND4 solar filters the extra-bright image can result in webcam exposures as short as 1/1000th of a second being achieved: ideal for freezing the seeing and capturing the finest solar details. As well as sunspots, a white light solar filter will record faculae (brighter, relatively white regions) and the solar limb darkening.

Telescopes and Focal Lengths for Solar Work

As we have seen, brightness and turbulence issues mean that large apertures are unnecessary in solar work. But what telescopes and image scales (arc-seconds per pixel) are the best? With a standard ToUcam Pro webcam, featuring 5.6 micron pixels, a focal length of around 2 meters is a sensible one to use, for sunspot imaging, in typical seeing. Such an arrangement gives an image scale of 0.6 arc-seconds per pixel, ideal for resolving fine sunspot details in white light. In good seeing, often experienced in the early morning, a few hours after sunrise, focal lengths of 3 meters or more can be tried, giving image scales finer than 0.4 arc-seconds per pixel. Rarely, if ever, is it worth sampling any finer than this. An ideal and relatively low-cost instrument for this type of work is a small apochromat (a perfectly color-corrected refractor) of 80- or 90-mm aperture (see Figure 16.1). With a 3× Barlow lens, or even a 4 or 5× TeleVue Powermate, such a modest instrument can

Figure 16.1. The solar imaging set-up of Damian Peach. An 80-mm Vixen apochromatic refractor and PC shielded from direct sunlight. Image: Damian Peach.

produce stunning high-resolution solar views with a webcam. Swap the webcam for a digital SLR and, say, a 2× Barlow and the whole Sun can often be bagged in one go.

As I type these words, the Sun is coming up to solar minimum and there are relatively few sunspots on the disc. However, large sunspots can appear at any time and small ones are fascinating to image with a webcam. In the days of solar photography, the "holy grail" was to photograph the "rice-grain" structure of the solar surface. In good seeing, the surface looks like it is made up of rice grains of just over an arc-second in size. In extraordinary seeing (resolving to 0.3 arc-seconds or finer with large apertures), the shapes of rice grains can even be recorded, and the solar surface looks like an area of "crazy-paving" seen from an aircraft. Even professional solar telescopes with vacuum-filled tubes mounted atop high towers struggle to better this resolution. But when they do, the resulting images are awesome.

Unlike the planets, there is no planetary rotation to worry about in solar imaging, but, having said that, there is less noise and more light, so there is not as much need to image for multiple minutes to stack thousands of frames. Noise is not a big issue here. Like the craters of the Moon, sunspots are a high-contrast feature and we are looking to just grab dozens of the very finest frames, rather than stack the maximum number. Also, as with imaging the moon, it is more obvious if large-scale seeing distortion occurs when you are imaging regions that may be five or more arc-minutes across. Registax stacks images by comparing features within the alignment box. Features outside that box, on a lunar or solar image may show serious distortions unless frames are selected with a high-quality threshold or are visually selected. A couple of excellent sunspot images, by Damian Peach, are shown in Figures 16.2 and 16.3.

Transits of Mercury and Venus

The occasions when amateur astronomers suddenly wish they had some expertise in solar imaging are those when Mercury and Venus transit the face of the Sun. Transits of Mercury are relatively common compared to those of Venus, although, with the kind of malicious cloud cover we get here in the U.K., an observer could conceivably miss every transit in his or her lifetime! When the transit of Venus occurred on June 8, 2004, it was the first Venus transit since December 6, 1882. The next transit will occur on June 6, 2012, but there is not another one until December 11, 2117. Amateur astronomers do not accumulate much experience in imaging Venus transits! Mercury transits occur at a rate of 13 or 14 per century. However, you do (obviously) need to be on the right hemisphere of the Earth, i.e., the daylight hemisphere, with the Sun above the horizon, to witness one. Mercury transits

JULY 17th, 2004. 16:02 UTC   Vixen 80mm FL APO @ F29. Baader Filter.
                                                                            D. Peach

Figure 16.2. A medium-sized sunspot on the lunar limb. July 17, 2004, 16:02 UT. 80-mm Vixen apochromat at f/29. ATiK 1HS webcam and filters. Baader solar film. Image: Damian Peach.

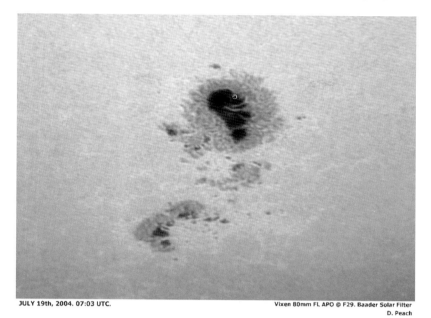

JULY 19th, 2004. 07:03 UTC.   Vixen 80mm FL APO @ F29. Baader Solar Filter
                                                                          D. Peach

Figure 16.3. A large sunspot on July 19, 2004, 07:03 UT. 80-mm Vixen apochromat at f/29. ATiK 1HS webcam and filters. Baader solar film. Image: Damian Peach.

can only occur in November or May. In my time as an amateur astronomer there have been five Mercury transits, but, due to geographic longitude and cloud I have only seen one: that of May 7, 2003. Amazingly, 13 months later, I saw the much rarer Venus transit, in a totally clear sky, from astronomer Patrick Moore's garden at Selsey in Sussex!

Needless to say, in the 122 years separating the 1882 and 2004 Venus transits, technology has moved on quite a bit. The 2003 Mercury and 2004 Venus events were the first to benefit from the application of webcam technology to the horrendous seeing conditions that result when the Sun is well above the horizon. Prior to this technological leap much had been made of the so-called "Black Drop" event seen at the 1882 Venus transit. Indeed, look at many astronomy textbooks of the 20th century and they will tell you that the black line joining Venus' limb to the solar edge at the 1882 transit was due to the Venusian atmosphere. In fact, webcam images of the 2004 event proved, categorically, that the Black Drop had nothing whatsoever to do with Venus, but plenty to do with poor seeing, poor instrumentation, and relatively long exposures. With perfectly collimated amateur telescopes and webcam exposures of 1/1,000th of a second (or less) the true nature of the Black Drop appears to have been resolved. In poor daytime seeing the disc of Venus seems to "breathe," i.e., to shrink and grow with the seeing. But when seeing is momentarily sharp, there was a clear gap between the black disc of Venus and the solar limb, even when only an arc-second of limb was left! In fact, the webcam results from the 2003 Mercury transit gave a clue that this would happen. Mercury too showed a Black Drop effect on frames exposed in moments of very poor seeing, despite that planet having no atmosphere. After the 2004 Venus transit images were analyzed it became obvious that the webcam had made a major scientific contribution. It had dispelled the 122-year-old myth of the Black Drop being caused by the Venusian atmosphere. A superb image of Venus transiting the Sun is shown in Figure 16.4.

Many excellent GIF animations were produced from webcam frames of the 2003 Mercury and 2004 Venus transits. With Venus' disc, almost an arc-minute across, taking 19 minutes just to cross the solar limb, these events are a leisurely affair, somewhat akin to watching paint dry. But at least there is time to correct technical problems, should they arrive. The 2003 and 2004 transits occurred during days when the Sun was virtually spotless, so the progress of the inky black dots across the solar disc was particularly dull. In the case of tiny Mercury, the disc size can vary considerably depending on whether the transit occurs in May or (three times as common) November. In May transits, Mercury's black silhouette can be 13 arc-seconds across, whereas in November transits it is a mere 10 arc-seconds. The next four Mercury transits (mid-transit times are given) are: November 8, 2006, 21:42 GMT; May 9, 2016, 14:59 GMT; November 11, 2019, 15:21 GMT; November 13, 2032, 08:55 GMT.

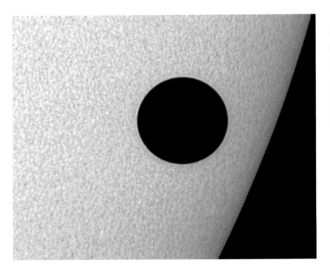

Figure 16.4. Venus leaving the solar disc after the rare transit of June 8, 2004 R+IR band (550–1000 nm). B&W Vesta Pro webcam. Astrophysics AP 130 f/6 apochromat + 2× Barlow and Herschel wedge. Image: Paolo Lazzarotti.

H-Alpha Viewing

In recent years there has been an upsurge in amateur astronomers who are imaging the Sun at H-Alpha wavelengths, i.e., at 656.28 nanometers. At this wavelength, the limb prominences can be seen, even while the blindingly brilliant Sun is in the field. Normally it takes a total solar eclipse to reveal these features. It should perhaps be explained here that the problem is not simply one of standard filtering.

The bandwidth of a good H-Alpha filter is, typically, less than one Angstrom (0.1 nanometers) and the filter production is highly complex and expensive. Decades ago Edwin Hirsch of the Daystar Company was the sole supplier of such narrow-band filters, but recently the Tucson-based company Coronado has been at the forefront of this technology and have developed a number of exciting products using advanced laser techniques. On the U.K.'s Isle of Man, Solarscope also offer quality H-Alpha filters of 50-mm aperture.

Both of these modern companies produce precisely tuned, ultranarrow linewidth classical Fabry-Perot air-spaced "etalons" for their H-Alpha filters. An etalon consists of a matched pair of ultrafine pitch polished and accurately figured fused silica plates. These have partially reflective and low absorption coatings for the desired transmission wavelength. To guarantee the essential fixed air space, the two etalon plates are skillfully assembled with the use of optically contacted spacers. Such filters have a very high throughput at peak resonance and a very narrow spectral transmission.

As one narrows the filter bandwidth centred on the 656.28-nanometer H-Alpha line the prominences become more and more sharp, and fine H-Alpha features on the disc emerge too. In the 1980s, the Baader Company advertised prominence telescopes in which a metal disc could be used with a telescope of a specific focal length to exactly occult the blinding solar disc. Using this method, even a 10 Angstrom H-Alpha filter would show the prominences, while the dazzling solar surface was hidden behind the metal disc. However, by moving to expensive, narrower bandwidth filters the prominences and subtle surface chromosphere features can be viewed simultaneously. Coronado makes filters and small quality refractors optimized for use with such filters. The 2005 Coronado range consists of H-Alpha telescopes from 40-mm to 90-mm aperture (ranging from $1,700 to $12,000 dollars in price) as well as individual filters priced from $900. These units typically have bandpasses less than 0.7 Angstroms. By stacking two matched H-Alpha filters together, a bandpass finer than 0.5 Angstroms can result! Recently, Coronado's wider, 1 Angstrom bandpass, 40-mm aperture f/10 PST (Personal Solar Telescope) has made H-Alpha imaging affordable to many and, coupled with a webcam, spectacular pictures of prominences can now be obtained for an outlay of under $500! The PST is mainly just a prominence telescope, and will show few fine details on the solar disc, but it is a remarkable price breakthrough. It may be thought that 40 mm is a very small aperture, but it is sufficient to resolve prominences only a few arc-seconds in width and perfectly compatible with typical daytime seeing. Like nearly all H-Alpha systems a filter tuning collar is provided to optimize the view and, in use it gives a sort of "3D" effect as tweaking it can enhance major disc detail or limb prominences.

H-Alpha Imaging

When you first look through an H-Alpha filter you will be struck by the deep red color of the image. It is immediately obvious that this is a redder red than you see in everyday life, and the color may be a bit off-putting at first. However, this is no problem for the CCD detector in a webcam, which is very sensitive at 656 nanometers. A monochrome webcam like the ATiK 1HS is the best choice as it is pointless using a color webcam for such narrow-band work. However, digital SLRs, and even small non-SLR digital cameras have been successfully employed and even hand-held to the eyepiece! The Scopetronix Company makes excellent digital camera-to-eyepiece adaptors. (www.scopetronix.com). Some early DSLR users have reported a moiré fringe effect when using an H-Alpha system, with the narrow wavelength somehow causing an optical interference with the camera's inbuilt filter and pixel grid but most users have reported no such problems. Some H-Alpha imagers have even used low-cost monochrome security cameras combined with a frame-grabber to collect images of prominences. However, using the same techniques as the planetary observer (already discussed at length in this book) will secure the best images, i.e., by using a monochrome webcam and processing in Registax. Of course, once you have the final stacked solar image there is nothing to stop you coloring it a more pleasant yellow/gold color in Adobe Photoshop or Paintshop Pro. In general, and especially in the cheaper, wider bandwidth filters, the solar disc is much brighter than the prominences and each may require a different degree of image processing in Registax. So, for whole disc images with a DSLR, the best solution is to use a longer exposure setting to record perfect prominences, but with an overexposed disc and a shorter exposure setting to just expose the disc features correctly. Each image can then be optimally processed and, with Photoshop or Paintshop Pro simply crop out the disc detail within the solar diameter from the shorter exposure image and paste it on top of the brighter prominence image. If using a digital camera you may well find that different color channels in the RGB image show different amounts of contrast, even though the image is supposed to be purely red! This is because digi-cam CCD pixel green and blue filters have some red leakage and, remarkably, the contrast in the green and blue channels can sometimes be more useful than the red!

However, for the very best high-resolution images, just use the same techniques as for the Moon and planets, i.e., a webcam, preferably monochrome, and Registax to process the AVIs. For more information on Coronado and Solarscope see the Appendix.

A digicam picture, by Maurice Gavin, using a Coronado PST, is shown in Figure 16.5. Maurice' Coronado PST itself is shown in Figure 16.6. Two superb H-Alpha images using a Pentax 75-mm refractor and a Coronado Solarmax 40 H-Alpha filter are shown in Figures 16.7 and 16.8.

Finally, whatever type of solar observing you do, do it safely. Let the camera and the webcam do the imaging and *not* your eye. Eyes cannot be replaced.

Good luck on your imaging adventure, whatever object in the solar system you choose to study!

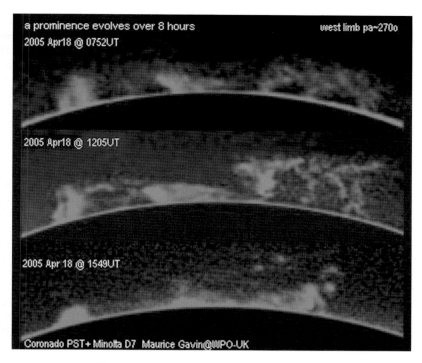

a prominence evolves over 8 hours   west limb pa~270o
2005 Apr18 @ 0752UT

2005 Apr18 @ 1205UT

2005 Apr 18 @ 1549UT

Coronado PST+ Minolta D7 Maurice Gavin@WPO-UK

Figure 16.5. An excellent montage showing the development of a solar prominence over a few hours, taken with a Coronado PST and Minolta digital camera. Image: Maurice Gavin.

Figure 16.6. The Coronado PST. A 40-mm aperture f/10 H-Alpha telescope for under $500! Image: Maurice Gavin.

Figure 16.7. An H-Alpha image of a solar prominence taken by Paolo Lazzarotti using a Pentax 75-mm refractor and a Coronado Solarmax40 H-Alpha filter. A Lumenera LU075M video camera was used. 300 out of 1,000 frames were stacked. Image: Paolo Lazzarotti.

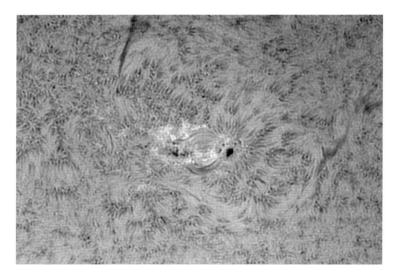

Figure 16.8. An H-Alpha image of a sunspot taken by Paolo Lazzarotti using a Pentax 75-mm refractor and a Coronado Solarmax40 H-Alpha filter. A Lumenera LU075M video camera was used. 500 out of 2,000 frames were stacked. Image: Paolo Lazzarotti.

Web Pages for Software, Hardware, and Observers

Web page addresses (URLs) change with time, especially those belonging to amateur astronomers. If a few of these links become obsolete during the production time of this book, simply go to a search engine like Google and type in the company name/amateur's name and a few key words. The new address should be located after a bit of searching.

Webcam/Video Software and Hardware

Registax Software by Cor Berrevoets: http://aberrator.astronomy.net/registax/
K3CCD Tools by Peter Katreniak: http://www.pk3.org/Astro or http://www.pk3.org/K3CCDTools
SAC cameras: http://www.sac-imaging.com/main.html
Celestron/Celestron NexImage: www.celestron.com/
IRIS Software: http://www.astrosurf.com/buil/us/iris/iris.htm
AVA, or Adirondack Video Astronomy (USA): www.astrovid.com
AVA's U.K. dealers True Technology Ltd: www.trutek-uk.com
ATiK webcams: http://www.atik-instruments.com/
ATiK UK Dealer: http://www.modernastronomy.com/
Astromeccanica (Italy): http://www.astromeccanica.it/ccd-camera.htm
Lumenera (USB 2.0 hi-spec cameras): http://www.lumenera.com/

Dr. Steve Wainwright's Quick Cam and Unconventional Imaging Astronomy Group (QCUIAG) is a mine of information on converting webcams and security cameras for astronomical use: http://www.astrabio.demon.co.uk/QCUIAG/

Software for converting a webcam to 'RAW' mode: http://www.astrosurf.com/astrobond/ebrawe.htm

Motorized Focusers, Barlows, and Powermates

Jims Mobile Inc. (Superb focusers and accessories): www.jimsmobile.com
TeleVue (the best Barlow lenses and Powermates): www.televue.com
TeleVue's U.K. dealer is Venturescope: www.telescopesales.co.uk

Safe Solar Filter Suppliers

Celestron: www.celestron.com
Kendrick: www.kendrick-ai.com
Orion USA: www.telescope.com
Tuthill: www.tuthillscopes.com
Thousand Oaks: www.thousandoaksoptical.com
Coronado H-Alpha equipment: http://www.coronadofilters.com/
Solarscope H-Alpha equipment (Isle of Man): Solarscope Ltd, Optical House, Ballasalla, Isle of Man, IM9 2AH.Tel: 01624 822724 Fax: 01624 620812. UK dealer is Venturescope: **www.telescopesales.co.uk**

Planetary Filter Suppliers

AVA/Adirondack Video Astronomy (USA): www.astrovid.com
True Technology (U.K.): www.trutek-uk.com

Quality Equipment for Planetary Observers

Cloudy Nights Equipment Reviews: www.cloudynights.com
Celestron: www.celestron.com
Celestron's U.K. dealer David Hinds: www.dhinds.co.uk
Orion Optics (U.K.): www.orionoptics.co.uk

Takahashi and Their Dealers
Takahashi Home Page (Japan) http://www.takahashijapan.com/
Texas Nautical (Takahashi U.S.A.): www.takahashiamerica.com
True Technology (Takahashi U.K.): www.trutek-uk.com

BC&F Telescope House (U.K.): http://www.telescopehouse.co.uk/
TEC (Telescope Engineering Company): http://www.telescopengineering.com/
TMB (Thomas M. Back) Optical http://www.tmboptical.com/

Software Bisque's Paramount ME mount www.bisque.com
Astrophysics (quality mounts and telescopes) www.astro-physics.com

CCD Manufacturers

SBIG: www.sbig.com
Starlight Xpress: www.starlight-xpress.co.uk
Apogee: www.ccd.com
FLI: www.fli-cam.com

Image Processing Software

Software Bisque (CCDSoft, The Sky, Orchestrate, and T-Point): www.bisque.com
Maxim (MaximDL/CCD): www.cyanogen.com
Richard Berry's AIP book and CD can be ordered from Willman Bell: www.willbell.com
AstroArt: www.msb-astroart.com

Planetarium Software

Starry Night: www.starrynight.com
Guide 8.0 by Project Pluto: www.projectpluto.com
The Sky from Software Bisque: www.bisque.com
Redshift 5: www.focusmm.co.uk
Skymap Pro: http://www.skymap.com

Optical Analysis Software

Aberrator Software by Cor Berrevoets: http://aberrator.astronomy.net/

Planetary Observing Organisations

British Astronomical Association: www.britastro.org/
Association of Lunar and Planetary Observers (U.S.A.): www.lpl.arizona.edu/alpo/
Association of Lunar and Planetary Observers (Japan): http://www.kk-system.co.jp/Alpo
The Astronomer Planets pages: www.theastronomer.org/planets.html
British Astronomical Association Mars Section: www.britastro.org/mars/
Communications in Mars Observations: www.mars.dti.ne.jp/~-cmo/oaa-mars.html
British Astronomical Association Jupiter Section: www.britastro.org/info/jupiter/

Hans-Joerg Mettig's JUPOS observations webpage: www.jupos.de
British Astronomical Association Saturn Section: www.britastro.org/info/saturn/
International Outer Planets Watch: http://atmos.nmsu.edu/ijw/ijw.html
International Marswatch 2003: http://elvis.rowan.edu/marswatch/
BAA Lunar Section recurrent TLP page: www.lpl.arizona.edu/~rhill/alpo/lunarstuff/ltp.html

Jet Stream and Weather Pages

Unisys Aviation weather and jet stream predictions: http://weather.unisys.com/aviation/
200-mb Jet Stream Forecast—Northern Hemisphere: http://grads.iges.org/pix/hemi.jet.html
European Jet Stream "Seeing": http://weather.unisys.com/aviation/6panel/avn-300-6panel-eur.html
Comprehensive Meteorological Data: http://pages.unibas.ch/geo/mcr/3d/meteo/index.htm
Weather chart archives: http://www.meteoliguria.it/archivio21.asp

Keen Planetary Observers

Damian Peach's excellent planetary imaging page: www.damianpeach.com
Antonio Cidadao's lunar and planetary imaging: http://www.astrosurf.com/cidadao/
Eric Ng's planetary images: www.ort.cuhk.edu.hk/ericng/webcam/
Ed Grafton's planetary page: www.ghgcorp.com/egrafton/
Dr. P. Clay Sherrod (Arkansas Sky Observatory): www.arksky.org/
Jesus R. Sanchez: www.arrakis.es/~stareye
Tan Wei Leong's planetary images: http://web.singnet.com.sg/~weileong/planets.html
Martin Mobberley's web pages: http://uk.geocities.com/martinmobberley/
Christophe Pellier: http://www.astrosurf.com/pellier/index2.htm
Thierry Legault's high-resolution imaging: http://perso.club-internet.fr/legault/
Christopher Go's planetary images: http://jupiter.cstoneind.com/
Mike Brown's lunar and planetary images: www.mikebrown.free-online.co.uk/

Spectacular Hubble/Spaceprobe Images

Hubble Space Telescope: http://oposite.stsci.edu/pubinfo/Pictures.html
Galileo Spacecraft: http://www.jpl.nasa.gov/galileo/
Cassini Spacecraft: http://www.jpl.nasa.gov/cassini/

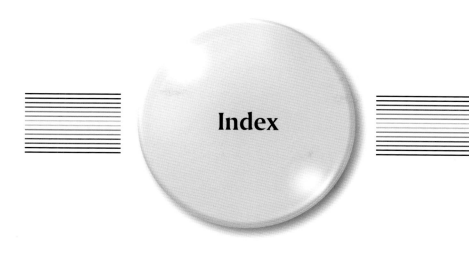

Index